The Lady Anatomist

PUBLISHED WITH THE
SUPPORT OF THE GETTY FOUNDATION

The Lady
Anatomist

The Life and Work of
Anna Morandi Manzolini

Rebecca
Messbarger

THE UNIVERSITY OF CHICAGO PRESS
Chicago and London

Rebecca Messbarger is associate professor in romance languages at Washington University, St. Louis. She is the author of *The Century of Women: Representations of Women in Eighteenth-Century Italian Public Discourse* and the coeditor and cotranslator of *The Contest for Knowledge*, the latter published by the University of Chicago Press in its series The Other Voice in Early Modern Europe.

The University of Chicago Press, Chicago 60637
The University of Chicago Press, Ltd., London
© 2010 by The University of Chicago
All rights reserved. Published 2010
Printed in China

19 18 17 16 15 14 13 12 11 10   1 2 3 4 5

ISBN-13: 978-0-226-52081-0 (cloth)
ISBN-10: 0-226-52081-1 (cloth)

Messbarger, Rebecca Marie.
The lady anatomist : the life and work of Anna
Morandi Manzolini / Rebecca Messbarger.
    p. cm.
Includes bibliographical references and index.
ISBN-13: 978-0-226-52081-0 (hardcover : alk. paper)
ISBN-10: 0-226-52081-1 (hardcover : alk.
paper) 1. Manzolini, Anna Morandi. 2.
Anatomists—Italy—Biography. 3. Wax-
modeling—Italy—History. I. Title.
QM16.M35M47 2010
611.0092—dc22
                                2009045488

In memory of my father
Paul Messbarger
and for
Audrey

Contents

Illustrations

Acknowledgments

Research for this study was supported by generous grants from the Mellon Foundation, the National Endowment for the Humanities, the American Philosophical Society, and Washington University. I am deeply grateful to my late father, Paul Messbarger, and to Wendy Roworth, Catherine Sama, Nina Gelbart, Seth Graebner, William Wallace, Walt Schalick, Inda Schaenan, and members of Washington University's Medical Humanities and Social Sciences Working Group for their helpful commentaries on drafts of parts of this book.

Over the years, I was provided a treasured forum and scholarly community for my work on Morandi by members of Washington University's Eighteenth-Century Studies Salon. I am especially grateful to Dr. Elio Vassena, who generously provided me with a copy of his thesis on the eighteenth-century history of anatomical wax modeling in Bologna; this proved to be a crucial aid to my search for archival materials relevant to the life and work of Anna Morandi. I thank Luciano Guerci, Paula Findlen, Gianna Pomata, Paola Giuli, Michael Shank, Adriano Prosperi, Simone Contardi, Marta Poggesi, Elizabeth Rodini, Phoebe Dent Weil, Romolo Dodi, Suzanne Magnanini, Tom Broman, Gianna Tomasina, and Christopher Johns for their assistance with important aspects of my research. I am profoundly indebted to Marta Cavazza for her nearly constant expert guidance during the course of writing this book.

I received crucial help with primary documents and their reproduction from Dr. Mario Fanti, superintendent of the Archiepiscopal Library of Bologna; Dr. Massimo Zini of Bologna's Institute of Sciences; Dr. Carmella Binchi, Dr. Maria Rosaria Celli Giorgini, and Dr. Anna Rosa Bambi of the Bologna State Archive; and curators Robin Rider and Micaela Sullivan-

Fowler of Special and Historical Collections at the University of Wisconsin. I am especially indebted to Dr. Fulvio Simoni, coordinator of the Poggi Museum, and other senior administrators of the museum, who contributed to this book by providing exceptional photos of the wax models in their collection. Mario Lorenzi, archivist, and Luciano Mingardi, custodian of San Procolo, kindly led me through the church, its archive, and its crypt. At the University of Bologna, professors Franco Ruggeri, Walter Tega, and Annarita Angelini facilitated my work with Morandi's wax collection, first in Via Irnerio and then at Palazzo Poggi. I was afforded the opportunity to present my research on Morandi to the Italian scholarly community by Raffaella Simili, the Association of University Women, and the International Center for the Histories of Universities at the University of Bologna. George Pepe was unstinting in his aid with translations of Latin texts.

I am grateful to Adam Kibel, MD, associate professor of urology at Washington University, who permitted me to observe a surgery involving the organs of the lower abdomen and assisted with the translation of remote anatomical terms in Morandi's notes on the male urogenital system. Neurosurgeon Andy Youkilis allowed me the extraordinary experience of observing as he operated on a spinal column and a brain. My concept of the human body and anatomical science was also transformed by Cheryl Caldwell, MD, Krikor Dikranian, MD, and the skilled group of anatomy professors and laboratory instructors who contributed to the gross anatomy course I attended at Washington University Medical School.

Daniel Garrison kindly provided copies of relevant excerpts of his translation-in-progress of Vesalius's *Fabrica*. I thank the art restorer Maricetta Parlatore for sharing her time and work on Morandi's collection with me and for providing photos of her restoration work. Lilla Vekerdy gave invaluable assistance throughout this project, locating rare archival materials and identifying obscure anatomical figures; and I am grateful as well to her always obliging colleagues in the Archives and Rare Books section of the Becker Medical Library at Washington University. I greatly appreciate the smart editorial assistance I received from Lee Spitzer and Courtney Weiss, as well as Rishi Rattan's capable technical and research support. Rita Kuehler facilitated innumerable aspects of my travel and research in Italy.

At the University of Chicago Press, I am profoundly grateful to my editor, Susan Bielstein, for her belief in this project and patience in seeing it through to print. Illustrations Editor Anthony Burton ensured that Morandi's story would come to life through the inclusion of relevant, high-quality images. Sandra Hazel edited the manuscript with tremendous skill and thoughtfulness. I am indebted to the anonymous readers of my manuscript for their enormously constructive and detailed suggestions for revisions.

My husband, Sam Fiorello, was my pillar and best critic. My sweet sons, Graham and Max, who are only slightly older than the book project itself, kept me focused on what is important. My mother, Pat Messbarger, by turns pushed and cheered me forward. I thank them and all my family members and friends, many of whom are mentioned above, for their loving encouragement through the course of creating this book.

....

Select sections of the book were revised and expanded from the following previous publications: "Waxing Poetic: Anna Morandi Manzolini's Anatomical Sculptures," *Configurations* 9 (2001): 65–97; "Cognizione corporale: La poetica anatomica di Anna Morandi Manzolini," in *Scienza a due voci*, ed. Raffaella Simili (Florence: Olschki Press, 2006), 39–61; "Re-membering a Body of Work: Anatomist and Anatomical Designer Anna Morandi Manzolini," *Studies in Eighteenth-Century Culture* 32 (2003): 123–54; and "As Who Dare Gaze the Sun: Anna Morandi Manzolini's Wax Anatomies of the Male Reproductive System and Genitalia," in *Italy's Eighteenth Century: Gender and Culture in the Age of the Grand Tour*, ed. Paula Findlen, Catherine Sama, and Wendy Roworth (Stanford, CA: Stanford University Press, 2009), 251–71.

FIGURE 1. Anna Morandi, self-portrait, wax, in process of restoration by Maricetta Parlatore; detail. Courtesy of Maricetta Parlatore.

FIGURE 2. Anna Morandi, self-portrait, wax, in process of restoration by Maricetta Parlatore; seen from back. Courtesy of Maricetta Parlatore.

Introduction

Changing the
Angle of Vision

Like perspectives, which rightly gazed upon Show nothing but
confusion eyed awry, Distinguish form.

SHAKESPEARE, *King Richard II*, 2.2.20–22

Standing alone in the cramped, makeshift restoration studio on the second
floor of the Department of Normal Human Anatomy at the University of
Bologna, I gazed down on my subject, the eighteenth-century Italian anatomist and anatomical wax modeler Anna Morandi Manzolini.[1] Morandi's
elaborate wax self-portrait stood on its base against the wall, literally split at
the seams. Cracks scarred her firm, round features, ears, arms, and fingers. Her
wig lay askew on her head; her elegant dress was discolored and frayed and her
straw insides spilled out the back (figs. 1–2). Morandi's current ignoble state
and her meager scrap-wood chambers secured by a dime-store padlock seemed
symbolic of her two-centuries-long depreciation in the annals of history. Yet
this would not be the site of her final un-making, but rather of her real and
conceptual re-membering.

The art restorer Maricetta Parlatore painstakingly scraped the cracks clean
of rotted wax; she blended resin and tinted paraffin to dress the wounds; and
with digital camera, needle and thread, straw and horsehair she put Anna
Morandi back together again. While Parlatore readied Morandi's facsimile for
the opening in 2000 of the restored museums of the Bolognese Institute of
Sciences in Palazzo Poggi, where her waxworks had been installed in 1776,
two years after her death, I labored to reconstruct a more complex and no less
evocative written portrait of the Enlightenment-era Lady Anatomist.

Re-vision

Anna Morandi surmounted meager origins, limited formal education, and endemic cultural and social constraints imposed on women and the working class to become the most acclaimed anatomical wax modeler of Bologna, at the time a European center for anatomical studies and wax anatomy sculpture. Among her patrons and admirers she counted Pope Benedict XIV, Emperor Joseph II of Austria, and Catherine the Great. Yet despite this impressive support, from her death two and a half centuries ago until the recent restoration of her waxworks, only a smattering of biographical sketches had acknowledged her contributions to eighteenth-century medicine and art. This book aims to redress the monumental discrepancy between the Anna Morandi of a tenuous, 250-year succession of biographers and the woman revealed in her vibrant, three-dimensional atlas of the human body.

A focal point of Morandi's plastic survey of human anatomy was the living eye and the intricate mechanics of sight, a leitmotif of the inquiry into the anatomy of perception she conducted throughout her career (fig. 3). Ironically, for contemporary leaders of Bolognese civic and intellectual life, she occupied a cultural blind spot. The value of her study of human anatomy and her art of anatomical wax design remained virtually invisible to local leaders—that is, until fervent outside approbation by European academies, dignitaries, and tourists brought it into view. Recognized by her elite compatriots for much of her career as neither anatomist nor artist, and her anatomical models as neither science nor art, the woman and her work dwelled all but imperceptibly in the hazy cultural interval saved for the Arts Mechanical. Since her lifetime, the same disciplinary and cultural liminality that diminished Morandi in the eyes of her Bolognese contemporaries has relegated her to historical obscurity.

Following the lead of her influential eighteenth-century biographer, Luigi Crespi, historians over the past 250 years have recurrently cast Morandi as a gifted dilettante, an anatomical *improviser*, and her rise in the field of anatomical science as both serendipitous and a corollary of her husband's career in anatomical wax design. She is portrayed as an untutored virtuoso whose talents spontaneously emerged at necessity's command. As helpmate to her melancholic spouse, she served him, according to these accounts, not merely in her role as wife but also as his dutiful assistant in the household anatomy laboratory and modeling studio he opened after battling with key members of the cultural establishment.[2] Upon his unexpected death, wifely duty and maternal care are again seen to have inspired Morandi to assume control of the household modeling studio and expand her anatomical practice. There was, of course, more to the story.

My reconstruction of Morandi's lifework has involved a systematic reevaluation of extant literature about her from the eighteenth century to the present

FIGURE 3. Anna Morandi, muscles of the eye, wax. Courtesy of Museo di Palazzo Poggi, Università di Bologna.

on the basis of primary documentation, while drawing on important recent historiography of the rise of anatomical wax modeling and the promotion of learned women in Enlightenment Bologna.[3] Archival documents, many of which came to light for the first time during my research for this book, not only provide such essential facts as the correct dates and circumstances of Morandi's birth and death, as well as the births and deaths of her husband and eight children; they also substantiate her expert knowledge and pivotal research in the field of anatomical science, even before her husband's death. And while municipal records and private correspondence testify to the challenges she faced as a working widow with two young children to support, they also reveal her keen ambition to achieve distinction in the field of anatomical science.

Documents tell only a partial story, however. This study necessarily discerns the woman and her lifework through the crucial lens of eighteenth-century Bolognese cultural history. It was an era of concerted, self-conscious renewal of the city's academic life: Bologna was largely defined by the establishment of a center of higher learning, the Institute of Sciences, which was independent of the university and dedicated to both the practice and the teaching of the "new" empirical science as well as the fine arts. Anna Morandi's meticulous, verisimilar anatomical waxworks represent a foremost reinterpretation—within Bologna and beyond—of the intersection of science and art established within the institute's walls.

Contours of a Body of Work

On 21 January 1714, in the parish of San Martino Maggiore at the heart of Bologna, Anna Morandi was born to Rosa Giovannini and Carlo Morandi.[4] On 13 March in the Palazzo Poggi, several blocks from Morandi's birth site, the Bolognese Senate inaugurated the Institute of Sciences (Istituto delle

FIGURE 4. Institute of Sciences, eighteenth-century engraving. Courtesy of the University of Bologna Library.

Scienze, fig. 4). These twin births produced two stories/histories that would shortly intersect. Each would contextualize and inform the other, as well as mark a transformative moment in Bolognese cultural history.

Beyond her birth date and her parents' names, almost nothing is known of Anna Morandi's first twenty-six years before her marriage to the Bolognese artist Giovanni Manzolini. No records tell us where she learned to read Latin or to write with eloquence and scientific exactitude, who oversaw her training as an artist, or how she came to know her husband.[5] While their paths undoubtedly crossed among the paint fumes and charcoal dust of a local art studio, her contemporary biographer and protector, the Bolognese senator Marcello Oretti, provides the sole primary reference to her early work as an artist, citing her "storiated paintings and excellent copies of the masters."[6] Of her familial relations, a single document confirms simply that at the time of her marriage she lived with her mother and twenty-three-year-old brother, Lazzaro, in the central parish of San Nicolò degli Albari.[7] Her early story thus draws predictably from the contemporary cultural milieu that cast and colored her life and her art of anatomical science.

Morandi was born into a Bologna determined to restore its prestige as the Mother of Learning (Madre degli Studi), faded after years of the university's decline. By the end of the seventeenth century, the number of students (especially foreign ones), the intellectual and economic lifeblood of the university and the city, had drastically decreased.[8] A damaging mix of patronage and politics had precipitated the fall: the imposition of an exclusively theoretical, as opposed to practical and experimental, model of teaching and the assignment by the Senate of nearly all university appointments to its own citizens, regardless of their qualifications or merit.[9] The archdeacon of the city, Anton Felice Marsili, voiced the distress of some members of the patriciate over the university's descent from cultural eminence and their demand for immediate reform:

> Let the wound be uncovered, and at this change of scene let the agonizing catastrophe emerge. One searches Bologna in vain for Bologna. The students disappear, the rostrums are empty, the teachers mute.[10]

It was none other than the archdeacon's brother, the outward-looking General Luigi Ferdinando Marsili, who would effect a radical cure to, in his words, "render this city, if not superior, at least equal to the major Modern Universities and Academies of France, England, Holland, and Germany."[11] In diametric opposition to the exclusively theoretical and oral teaching at the university, Marsili founded the Institute of Sciences, in whose specially equipped halls and laboratories an elite corps of publicly funded professors exercised the tripartite method of "experiment, classification, and exposition" in their research and teaching (fig. 5).[12] And while each of the professors occupied an assigned post within a recognized discipline, the architectural distribution of the laboratory and workspaces in the institute aimed to optimize collaborative exchange among them. This interchange, as Marsili ideally conceived it, would not only bring together the body of scientific investigators in chemistry, physics,

FIGURE 5. Antonio Zanca and Antonio Calza, General Luigi Ferdinando Marsili. Museo Marsiliano, University of Bologna Library. Courtesy of Museo di Palazzo Poggi, Università di Bologna.

natural history, and the like, but would bridge the enduring divide between the sciences and the fine arts. The institute united under Palazzo Poggi's one roof the Academy of Science, formerly of the Inquieti, and the Clementina Academy of Art.

From the start, the Clementina was a markedly more conservative society than the institute's Academy of Science, vigorously endorsing Bolognese artistic tradition against innovation. As the historian Anton Boschloo has observed, its chief mission was

> the maintenance and transmission . . . of the "inheritance of the great Bolognese painting of the seventeenth century" . . . ; the defense of the Bolognese artistic patrimony; and the reinforcement of the figurative arts' position within the liberal arts, in opposition to their lower status as professional associations or crafts.[13]

Notwithstanding the backward-looking cultural disposition of the Art Academy in contrast to the progressive Academy of Science, the union of these two entities within the same Institute of Sciences reveals a vision of the inherent attachment and overlap of science with art that was a defining, albeit controversial, attribute of eighteenth-century Bolognese intellectual life and practice.

Marsili died in 1730, when Anna Morandi was only sixteen years old and still more than a decade away from her career in anatomy and anatomical design. However, the principal intellectual and pedagogical ideals he had established within his *House of Science*[14] continued to underpin scientific inquiry and methods when she took up the scalpel and colored wax, and they unmistakably guided her anatomical practice. Morandi would implicitly espouse Marsili's approach of the *naturalista metodico* (methodical/systematic natural philosopher), his tripartite method of "exposition, classification, and experiment"—or, to use the eighteenth-century Italian term, *esperienza* (experience)—to uncover the intrinsic order and underlying operations of the human body, truths of practical use to the individual and the world.[15] She would convey these truths, moreover, through her novel art of science, an ideal merger between anatomical experiment and classification, and artistic exposition that Marsili could not have conceived.

The Enlightenment Pope

Marsili's most important ally and successor in the campaign to restore Bologna's cultural distinction was Prospero Lambertini (1675–1758), the influential Bolognese nobleman, cleric, and future Pope Benedict XIV, from whom Morandi would come to receive crucial support at a key juncture in her anatomical career (fig. 6). In 1726, the Holy See dispatched him from his post as custodian of the Vatican Library to arbitrate a conflict between Marsili and the Bolognese Senate over the management of the general's bequests to the institute as well as the patronage-based hiring practices for the institute enforced by the Senate.[16] Lambertini sanctioned scientific over political interests in the dispute and thus began a lifelong advocacy for the institute and the academic prestige of his birthplace.[17] When he returned to Bologna as cardinal archbishop in 1730, the year of Marsili's death, Lambertini launched a campaign to expand the ongoing reform of the academic life of the city. Given the intractable conservatism of the university, he deemed the institute the key means for reestablishing Bologna as a premier center of learning in Europe: "If God will help us, and before dying we might give it a bit of aid, the Institute is capable of making our homeland famous, as in other times the University was famous."[18]

More than any other scientific discipline, Lambertini considered medical science, especially the study of anatomy, the wellspring of Bologna's past glory and future cultural ascendancy. From his homecoming as archbishop through his rise to the papal throne, he sought to revive the city's legacy in anatomy built by such masters as Mondino de Liuzzi, author of the most important medieval treatise on practical anatomy; Marcello Malpighi, the founder of microscopic anatomy; Malpighi's student Anton Maria Valsalva, famous for his investigations of the anatomy of the ear; and Valsalva's student Giovan

FIGURE 6. Giuseppe Maria Crespi, *Pope Benedict XIV*. Pinacoteca, Vatican Museums, Vatican State. Photo: Scala / Art Resource, NY. Used by permission.

Battista Morgagni, father of anatomical pathology. Lambertini established the first chair of surgery at the Institute of Sciences for the celebrated young anatomist Pier Paolo Molinelli, whom the archbishop charged with conducting anatomy lessons and dissections in the city's two main hospitals.[19]

As archbishop, Lambertini wrote several important decrees for the practical advancement of modern anatomical studies. In particular, his *Notificazione* of 8 January 1737, "On the Anatomy to Be Done in Public Schools," demonstrated the intensity of his support for new scientific methods in the study and practice of human anatomy.[20] Lambertini reinterpreted Pope Boniface VIII's bull *Detestande feritatis* of 1299, which had condemned on pain of excommunication anyone who exhumed, sectioned, or boiled the bodies of buried Christians.[21] According to the archbishop, Pope Boniface had never intended for his order to apply to the necessary and publicly useful anatomical dissection practiced at the universities. At the conclusion of the *Notificazione*, he issued an extraordinary directive to local clergy to urge Bologna's faithful to donate their dead kin for the purpose of anatomical research:

> As regards the cadavers not of the condemned . . . but rather the cadavers of men or women . . . dead by whatever other means, that are believed to be necessary for the study of Anatomy taught at the University, let a petition be made either to Us or to Our General Vicar, with assurance that, in order to facilitate this useful work, all opportune measures will be taken for the consent of the Relatives, for the authorization of the Parish priest, and for the Funeral Services.[22]

Notwithstanding Prospero Lambertini's authentic support of the "new" science of anatomy in Bologna, it was his erection of an anatomy museum in the heart of the Institute of Sciences for the display of spectacular, life-size wax anatomical figures that would epitomize his patronage of human anatomy as well as his stage management of Bologna's cultural reform. Palpably reminiscent of the criminal bodies ritually dissected at the annual Public Anatomy held during Carnival in the University's Archginnasio Anatomy Theater, the museum's figures, realistically cast in colored wax, were a dramatic monument to Bologna's authority in anatomical science (fig. 7). Much like the annual Public Anatomy, however, the Anatomy Museum actually did not serve the specialized study and practice of the science itself. Indeed, its anatomical wax figures belonged more to art than to science. The museum's focal series of four life-size écorchés, or flayed men, expressed in scientifically precise and aesthetically classical terms the anatomy of greatest import to sculptors and painters: the muscles and bones that move the body. The artist and anatomical wax modeler Ercole Lelli, in fact, explicitly championed their use by Bolognese artists as a superior paradigm of anatomical design. Yet, while Lambertini understood their importance for more accurate representations of the body in

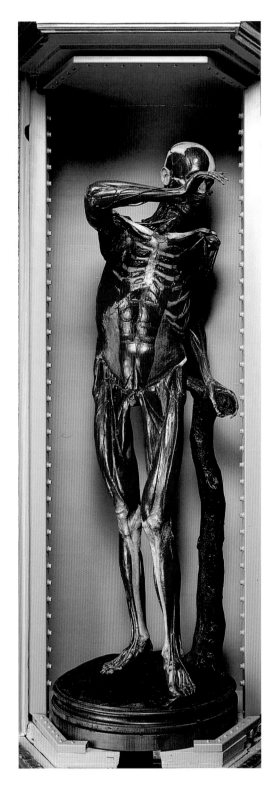

FIGURE 7. Ercole Lelli, superficial muscles, wax and bone. Courtesy of Museo di Palazzo Poggi, Università di Bologna.

art, he viewed the wax anatomies above all as an effective means of attracting to his native city the attention and admiration of Enlightenment Europe.

Most important for our purposes, however, the pope's museum was the accidental font and lasting foil of Anna Morandi's very different art and science of human anatomy. As will be seen, in diametric opposition to Lelli's conception, Morandi created her collection for use by medical professionals and avid amateurs in the study of practical anatomy. Her wax bodies supplanted and indeed surpassed in critical ways real dissected bodies, whose defunct systems, parts, and attachments she could purify of their natural ambiguities and, using vibrant colors and three-dimensional movement, imbue with virtual life.

The Making of an Anatomist

Chapter I tells the story of the pope's museum as a frame for the origins and distinctions of Anna Morandi's anatomical practice. Of central importance, late in 1746, Giovanni Manzolini resigned after three years as chief assistant on the papal commission for the Anatomy Museum. A swell of bitter resentment ruptured Manzolini's relations with the project director, Ercole Lelli, who, Manzolini believed, had robbed him of deserved acclaim for his superior proficiency in anatomy and anatomical sculpture. With his artist wife, Anna Morandi, Manzolini opened in his home a rival wax-modeling studio and school of anatomy for the instruction of medical students and keen amateurs. So it was that Anna Morandi, the thirty-two-year-old wife, mother of two young sons, and professionally trained artist, entered a critical sphere of eighteenth-century Bolognese cultural life on the borderland between art and science. Morandi's study and representation in wax of human anatomy thus arose directly from the pope's museum project; but as will be seen, the comprehensive scope, functional exactitude, and subdued aesthetic of her representations contrasted diametrically with the moralizing ritual and spectacular science of the annual Public Anatomy recollected in the pope's museum. The challenge she posed to the official representation and use of anatomical wax models would consign her waxworks to the margins of eighteenth-century Bolognese culture.

The characteristic features of Morandi's early anatomical practice with her husband are the focus of chapter 2. For nearly ten years, Anna Morandi stood scalpel to forceps with her partner-husband, dissecting hundreds of cadavers in their household anatomy school and laboratory. Although they studied all the parts and systems of the body, I will show that they specialized in some of the age's most innovative areas of anatomy, including the reproductive and the sensory organs. The Bolognese obstetrician and professor of surgery Giovan Antonio Galli commissioned the couple to sculpt the models of the gravid uterus and the female reproductive system for Bologna's first school of obstetrics, which he opened circa 1749 in his home;[23] the school was later installed at Pope Benedict XIV's command within the Institute of Sciences.[24] They also created for Bologna's first chair of practical surgery, Pier Paolo Molinelli, pioneering obstetrical models as well as a special series of models of the ear.[25] Beyond commissions from leading professors of medicine in their native city, the couple received orders from academies throughout Italy and Europe for their exacting replicas of the anatomical body.

While Anna Morandi and Giovanni Manzolini worked together dismembering real bodies and ideally re-membering them by means of plaster casts and colored wax, she alone was the public face of their home studio. As described in numerous published and private primary accounts, Morandi, never her husband, gave regular lessons and hands-on demonstrations of their vast collection of anatomical wax models to medical practitioners and local and foreign visitors, who increasingly flocked to the studio. Morandi's hands-on fluency in anatomy as well as her public authority and refinement in these demonstrations were objects of fascination and praise for members of the elite academic community and a range of eminent tourists.

Chapter 3 juxtaposes the empirical and scholarly anatomical studies Morandi conducted in partnership with her husband as well as the eyewitness accounts of her anatomy demonstrations recorded by visitors to their studio with conventional biographical sketches of her as a "female improviser" in anatomy inspired by wifely duty and innate genius. Much like the representation of famous female improvisational poets of her age, Morandi's feminine body and sensibilities were seen as essential to her art of dismantling and reproducing the anatomical body. As I aim to show, the characterization of her entry and work in anatomy as dispossessed of any but her instinctive abilities was first put forth for ideological reasons by her contemporary biographer, Luigi Crespi, in 1769, but has refigured in biographical sketches from her lifetime to the present. As a counterpoint to stock representations of her, this chapter features Anna Morandi's own voice of authority conveyed in letters defending the anatomical practice she shared with her husband.

Morandi's Visual and Literary Oeuvre

While direct testimonials authenticate Morandi's role in Enlightenment Bologna and the nature and extent of her influence in the field of anatomical science, the most important primary sources are, of course, the ones she authored. These are the focus of analysis of the second half of the book.

With the unexpected death of her husband in 1755, the thirty-nine-year-old anatomist and artist assumed control of the household wax-modeling studio. She continued in the anatomical practice she and Giovanni Manzolini had built over the course of a decade even as she pioneered several notable improvements. On her own, she invented superior wax compounds and methods of sculpture that rendered her waxworks more resistant to wear and allowed her to show traits never before represented to the naked eye. She also developed more precise techniques of human dissection that led to her discovery of select new components of the body. Perhaps most important, Morandi composed a detailed, 250-page anatomical notebook that indexed most of the body parts that she had conceived in wax, explaining their structure and physiology, describing her anatomical method, and critically engaging dominant anatomical theories. Morandi's schematic narrative of her work in anatomy and anatomical design, which until very recently has been virtually disregarded as a subject of analysis in extant biographical sketches, is a focal point of this book.[26]

Her sculptures and extensive explanatory notes formed a single narrative, visual and literary, of her study of human anatomy. This compound oeuvre manifested the intimate partnership between the systematizing eye and the probing hand that marked the Enlightenment worldview. Bridging what Barbara Stafford has judged the "untraversable abyss of the eighteenth century between the practical visual and the theoretical textual,"[27] her works conjoined the fine arts and surgery, sense and cognition, hands and eyes, word and image, to theorize and envision the workings of the human body. Her sculptures and voluminous explanatory notes offered a rendering of the *fabric of the human body* that conceived each organ in terms of its vital function within the context of a dynamic, interdependent, physiological whole.

My reading of this narrative over four chapters considers its range of meaning within the context of Enlightenment Bologna and of the history of science and medicine. I explicate Morandi's corpus of the body in order to define her method of scientific inquiry as well as the origins and development of her theories of anatomy and physiology. The extant catalogue of Morandi's personal archive of anatomical and medical texts and her frequent written references to authorities, ancient to contemporary, in the science serve as crucial maps of both the theoretical and the visual foundations of her work.[28]

As mentioned at the outset, Morandi's story is puzzled together from a thin primary record and a densely detailed Bolognese Enlightenment context. Consequently, my interpretation of Morandi's lifework is at its core an interdisciplinary cultural project. Among numerous relevant topics, I consider the influence of Bologna's networks and politics of patronage on Morandi's practice of anatomy. I examine her achievements in light of the remarkable authority granted scientific women in Enlightenment Bologna, and I aim to explicate the vigorous polemic in Settecento Italy about the education of women, a polemic in which Morandi was directly implicated. I also view Morandi's work and cultural status in terms of the gristly *stufato*, on which the Bolognese academic class gnawed for years, over the purpose and scope of anatomical design. Not surprisingly, this fierce debate hinged on class divisions between art and craft, artist and artisan.

Self-made

While Morandi's body unmistakably inhered in her wax anatomies (her collection doubtless included models cast from her own hands and face), she made herself the explicit subject of contemplation in her extravagant, life-size, wax self-portrait. In chapter 4, my interpretation of Morandi's oeuvre begins with this imposing, tactile title page to her three-dimensional anatomical atlas. Literally casting herself as Bologna's Lady Anatomist, she is shown in the act of dissecting a human brain while improbably arrayed in the elegant dress and accoutrement of a noblewoman. Through this visual autobiography, she simultaneously sublimated her social status and authority in the field of anatomical science even as she repressed all signs of her work as artist and wax modeler. As I aim to show in this chapter, her self-portrait evoked and at the same time rivaled the iconic image of the Renaissance master anatomist Andreas Vesalius on the title page of his revered *Fabrica*. It is from this commanding, indeed defiant, image that the title and viewpoint of my book originate. I have deliberately sought to bring Morandi's self-representation as the Lady Anatomist into focus over lesser portraits of her created during her lifetime—not to exaggerate her influence but instead to emphasize the distinction between the way she conceived herself, indeed the way the world beyond Bologna's gates saw her, and the minor part assigned her at home and in history.

Although her skill in anatomical wax design was known in academic centers throughout Italy and Europe before her husband's death, Morandi inspired greater interest after being widowed. As a woman alone dissecting cadavers and demonstrating her wax studies of anatomy to crowds in her home studio, her exceptional status rose by many degrees. Italian and European courts and academies also hoped that she might now be willing to leave Bologna. She received invitations from throughout Europe to relocate her practice and singular collection. It was precisely this outside notice and the potential risk of losing such a popular Grand Tour attraction that prompted Bologna's cultural patriarchs to recognize her work formally with a small annual stipend and a university appointment in anatomical exposition. Pope Benedict intervened with the Senate in these matters in Morandi's behalf, thereby keeping her home in Bologna.

Morandi was at this time also the unmistakable beneficiary of a distinctly liberal view of learned women in Bologna, a view cultivated by Prospero Lambertini during his term as archbishop and afterward as Benedict XIV from his papal throne in Rome. As Paula Findlen and Marta Cavazza have demonstrated, learned Bolognese women were vaunted emblems of the city's cultural resurgence and celebrated proof of its incomparability as a nexus of Enlightenment progress and social and intellectual exchange.[29] In ways reminiscent of his backing of Laura Bassi, the first woman to receive a degree from the University of Bologna, the pope defended the symbolic and material value to Bologna of being home to the Lady Anatomist.

Morandi's Anatomical Atlas

Despite the loss of numerous wax models and notable gaps in the documentation, it is clear that Morandi's oeuvre encompassed most of the organs and parts of the body. Featured among her waxworks were figures intact and anatomized, in some cases to the microscopic level, of the heart, the forearm, the leg and foot, the larynx and pharynx, the muscles and bones of the face, the mouth, the jaw and tongue, the ear and eye shown in normal and colossal proportions, the hand, the respiratory apparatus, the kidneys, the hollow and gravid uterus, and parts of the male urogenital system. Extant waxworks created with her husband or on her own encompassed portions or all of the following: the muscular system, the skeletal system, the nervous system, the circulatory system, the respiratory system, the urinary system, and the male and female reproductive systems. Morandi's anatomical notebook, which served as a teaching text for the lessons she gave to medical students and as an inventory and explanation of portions of her work for her patrons, detail her series on the pharynx and the larynx; the leg and forearm created together with her husband; and her collection of intact skeletons and discrete bones ranging from fetal to adult development. However, the leitmotifs of her written oeuvre and the subjects most celebrated by visitors to her studio were the sensory and the reproductive organs. As one might imagine, the latter provoked some rather piquant comments.

Chapters 5 and 6 focus respectively on Morandi's study of the sense and the sex organs. She devoted 110 pages of her 250-page notebook and a vast though unknown number of wax models to the study of the eye, ear, nose, tongue, and hand. Despite the mysterious loss of nearly all twenty-two of her wax models of the male reproductive system and genitalia, her 45 pages of meticulous notes on the subject as well as the high price she received for the

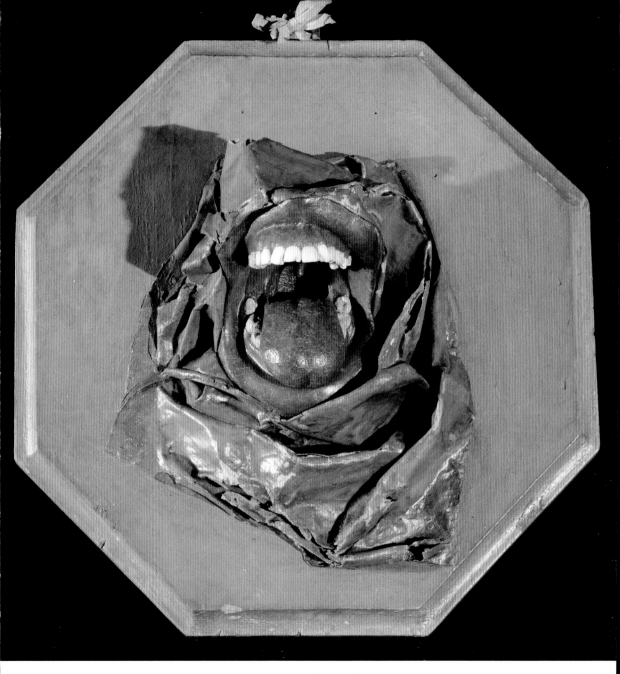

FIGURE 8. Anna Morandi, mouth and tongue, wax. Courtesy of Museo di Palazzo Poggi, Università di Bologna.

series (ten times that of any other part of her oeuvre) are proof of the special importance and fascination it held for eighteenth-century viewers.

The Art of Anatomy

The three-dimensional bodies that Morandi cast in wax existed in rapport and congregate with the world that surrounded them. Extant illustrated organs, limbs, and physiological systems reach, often literally, beyond the borders of their display tables, evoking both the whole living body of which they are a part as well as the concrete realm of human experience. Expressive flayed faces look about, engaging the gaze of their viewers, active skeletal arms stretch and grip, tongues jut forth from open mouths (fig. 8), and a placenta spilling from a distended uterus beyond the edge of its display table conveys the surge of childbirth into the space of the living. While classical landscapes replete with Roman ruins and intricately drawn flora traditionally framed the flayed and eviscerated subjects of the early modern anatomical atlas, Morandi's waxen atlas instead occupied the domain of actual human experience and exchange. As I strive to show in chapter 5, this appeared most poignantly in her manifold representations of the organs of sense, specifically her renderings of the hand and the eye, those parts indispensable for anatomical dissection and sculpture, sense and cognition, perception and knowledge.

Beyond an examination of Morandi's distinct science of anatomy, her method, authorities, and object as anatomist and modeler, I consider the epistemological foundations of her oeuvre. Drawing on critical analysis of the body and corporeal metaphors in early modern visual and theoretical texts, my reading thus extends to the poetics of her body narrative, the transubstantiation of wax limbs and organs into the wondrous order and workings of life-matter, being and knowing, and the objects and intersections of art and science. Morandi's own body was, of course, seen (indeed frequently looked upon) by her contemporaries as the uncommon vehicle of her unique art of anatomy. It was, in fact, through her hands-on knowledge and re-production of bodies that she took epistemological control of a conventional male system of scientific discovery and knowledge. I analyze the ways in which Morandi's mastery of anatomy engendered essential revisions of the science, its ideal method and practitioners, and the relationship of science to art, and art to craft.

Of obvious importance, therefore, are questions of sex and gender as they relate to Morandi's career as a female anatomist/artist, and as they elucidate her prolific representations of the male and female anatomy. Taking as a point of departure in chapter 6 feminist critiques of the objectification of women during the eighteenth century by male anatomists and anatomical designers, I distinguish the ways in which Morandi disrupts the patrilineal hold on the body by upending the tradition of the male scientific observer and the female anatomical subject. I also view Morandi's comprehensive study of the male reproductive system in light of the ongoing academic controversy over Thomas Laqueur's theory of the emergence during the eighteenth-century of a two-sex model.[30] The unorthodoxy of Morandi's role as a woman anatomist, which excited both strong approval and invective from Bologna's *professori*, her unabashed claims to expertise in the male anatomy, and her implicit refusal to conceive of the reproductive human body in terms of the conventional male/female opposition, serve to unsettle and extend in significant ways current theories about the meaning of sexual difference in the eighteenth century.

The final chapter of this book carries Morandi's story, and what I contend was the heyday of the anatomical wax modeler, to their common close. Morandi spent the final years of her life as a celebrated guest resident in the palace of Senator Count Girolamo Ranuzzi, who purchased her complete oeuvre and played host to eminent visitors to the Lady Anatomist as a means of augmenting his own stature and influence. Analysis focuses on the effect of Ranuzzi's patronage on the reception of Morandi's art of anatomy, before and after her death in 1774. In 1776, Morandi's oeuvre was purchased from the count by the Senate of Bologna and installed in its own room in the Institute of Sciences. This posthumous civic recognition of Morandi's mastery of the science and the art of anatomy ironically signaled the culmination of her influence and the concomitant fall of the anatomical modeler. As I will discuss in this final chapter, the center of production of anatomical wax models shifted at the precise moment of Morandi's death from Bologna to Florence, where in 1775 the Royal Museum of Natural History opened its doors. The centerpiece of the museum was a collection of hundreds of graphic anatomical wax models conceived by the famed natural philosopher Felice Fontana but created by an assembly line of "manual labors," from sextons to barber surgeons and anatomists, who prepared countless cadavers for casting, to numerous "artisans,"[31] who molded and sculpted the wax models. Although Fontana had studied in Bologna and taken his inspiration from the work of Ercole Lelli and Anna Morandi, as Maerker has shown he explicitly referred to the anatomical modelers in his charge as his "tools," and viewed their specialized contributions to the museum project as far beneath his own and often with open contempt. From this point, therefore, the manual and mental labor, the art and the science, of anatomical wax modeling would no longer be the province of a single genius, but would instead be divided according to a rigid hierarchy of knowers and makers.[32]

Partial View

Of the extant wax models sculpted by Morandi, a significant number were carefully restored in 2000 for permanent exhibition in the refurbished Palazzo Poggi, the original site of the Bolognese Institute of Sciences. However, numerous sculptures, especially those commissioned by foreign patrons, have not been recovered, and others, in particular her collection of skeletons ranging from fetal to adult development, remain out of sight in extreme disrepair. In composing this account I have endeavored to narrow lacunae in her story when possible, but also to indicate those gaps that have not yet been filled. Primary texts, ranging from the recently unearthed letter from Catherine the Great in praise of Morandi to her elaborate six-foot tombstone installed at her death in the central apse of San Procolo but which has been hidden for more than a century in the church's crypt, serve not only to recapture Morandi's attainments but also to illumine her story's near unraveling from the historical record.

This book is thus also about the partiality of Morandi's story. As she evoked the whole, living person through her sculptures of incomplete faces, solitary eyeballs, and truncated hands, I hope to render the unique force and trajectory of her life and work despite fragmentary biographical data and the loss or damage of critical parts of her oeuvre.

New Views

Since I began researching and publishing on the lifework of Anna Morandi, several important studies have appeared to fill out and enrich her story. Of particular merit is the well-documented article by Gabriella Berti Logan, "Women and the Practice and Teaching of Medicine," which examines the influence of scientific women, including Morandi, in Bologna during the eighteenth century. Lucia Dacome's several essays of note on eighteenth-century Italian anatomical wax modeling and the waxworks of Giovanni Manzolini and Anna Morandi illuminate the cultural context as well as the performative and, intriguingly, the "alchemical" aspects of the craft.[33] Miriam Focaccia's essay, which serves to introduce the first publication of Morandi's anatomical notebook, recapitulates extant scholarship on Morandi and related subjects, while adding crucial new facts to her biography.[34] Also noteworthy are the rigorous studies by Simone Contardi and Anna Maerker of the history and cultural significance of the anatomical wax models created for the Royal Museum of Physics and Natural History (the Specola) in Florence. Florence, of course, drew its inspiration from and essentially replaced Bologna as a center of anatomical wax production immediately after Morandi's death.[35]

Gazing on the Gazer

Anna Morandi looked where others, most especially other women, rarely dared. She entered that most virile of the "new" sciences, keeping company with the dead, handling cadavers and "fresh parts" carted to her home from the city mortuary. And in her three-dimensional wax translations she reinscribed with robust vitality and elegant wholeness the lifeless, putrefying flesh unstitched and explored with her dissecting scalpel and her critical eye. Her work at once informed, intrigued, titillated, and repulsed her viewers, and for that reason drew a vast public from throughout Italy, Europe, and beyond. She was for some a menacing interloper, whose innately inferior female body and immodest gaze diminished the authority of anatomical science and its properly male masters. Her most vociferous opponent was the Bolognese professor of anatomy Petronio Zecchini, whose acid denouncement of her work served as preface to his 1771 tract theorizing that the cause of women's inherent intellectual deficiency was their reproductive organs, what he terms their "thinking uterus."[36] For others she symbolized instead the Mother of Anatomy. The noted physician and botanist Giovanni Bianchi of Rimini underscores, in the published account of his visit to Morandi's studio in September 1754, the power of her maternity to set her apart from and indeed above the society of male anatomists.[37] Notwithstanding the extreme converse views expressed by Zecchini and Bianchi, the Lady Anatomist captivated her contemporaries. It is my hope that her lifework and the distinctive alliance she formed between numerous worlds and conceptual fields in her study of human anatomy will rouse the interest and the imagination of readers in the twenty-first century and beyond.

I

The Pope's Anatomy Museum

If in perfection tempered were the wax,
And were the heaven in its supremest virtue,
The brilliance of the seal would all appear;

But nature gives it evermore deficient,
In the like manner working as the artist,
Who has the skill of art and hand that trembles.

DANTE, *Paradise*, canto 13

A man, young lady! Lady, such a man
As all the world—why, he's a man of wax.

SHAKESPEARE, *Romeo and Juliet*, 1.3.75–76

At the confluence of two rivers of ambition, Bologna in the eighteenth century was uniquely suited to become the European center for modeling the anatomized human body in wax. Archbishop Prospero Lambertini's resolve to advance his native city's reputation as a premier site of science, and the Bolognese artist Ercole Lelli's (1702–1766) mastery of anatomical sculpture and fierce hunger for a leading role in Bologna's cultural revival coalesced to elevate anatomical wax sculpture to a spectacular sign of the *new* Bologna (figs. 9–10).[1] The Bolognese nobleman and future Pope Benedict XIV blessed with his enduring favor and financing the artist Lelli, whose wax anatomies would ally art, science, and religion through a novel integration of a practical didactics of the body with a spectacular and solemn aesthetics. A dramatic emblem of the communion between faith and the *new* science espoused by the Enlightenment pope, the Anatomy Museum that he would establish within the Institute of Sciences simultaneously evoked sacred and scientific spaces and aims.[2] It

FIGURE 9. Pierre Hubert de Subleyras, *Pope Benedict XIV*. Pinacoteca Nazionale, Ferrara, Italy. Photo: Finsiel/Alinari / Art Resource, NY. Used by permission.

conjured church and Public Anatomy theater, reliquary and natural history cabinet in its graphic and moralizing display of the anatomized body.[3]

Ercole Lelli's anatomical wax figures mingled a neoclassical heroism and a "dust to dust" morality with the scalpel's precision. He proffered through his anatomical sculptures a rebalancing of the retributive ethic and spectacular aesthetic of the baroque memento mori with empirical scientific practice.[4] Quite simply, his anatomically precise, pathos-laden anatomical figures epitomized the hybrid drama of sin and science played out annually at the university's renowned Public Anatomy, or Carnival Dissection, as it was known. Placed by the pope in the prestigious Institute of Sciences at the intersection between the Academy of Science and the Clementina Academy of Art, Lelli's series of life-size anatomical figures served to dramatize Bologna's historic and renewed distinction in anatomical science while at the same time exemplifying for Bolognese artists a more exacting standard of anatomical design. The waxworks specifically demonstrated the anatomical apparatuses of muscle and bone that move the body, those parts most important to painters and sculptors. On a practical level, with Lambertini's prominent backing, the anatomical waxes represented a specialized new art form.[5] But they did not in fact serve the practical science of anatomy itself.

The Anatomy Museum was above all an ideal means for riveting international attention on La Dotta, Bologna's lofty ancient moniker meaning "The Learned," that Lambertini clearly favored over the more familiar body-bound epithet La Grassa—"The Fat." By placing on permanent display a graphic

ous Carne Vale! or Farewell to Meat! until the Easter thaw. Raucous festivities, permissive masquerades, and all varieties of bodily indulgence upturned rigid social codes in anticipation of Lenten penitence and deprivations. It was at this time that the performance of the annual Public Anatomy, otherwise known as Carnival Dissection—a very different kind of *farewell to meat*—took place in choreographed stages during the fifteen days before Ash Wednesday.

A series of official acts traditionally anticipated the Public Anatomy. The capital execution of a local criminal was an elaborate public ritual that served as prologue to his dissection, which commonly occurred after sunset on the same day.[12] In his redolent analysis of what he calls the "liturgy of public executions," Lionello Puppi unravels the intricate regulations and the moral and political logic of "the highest, most difficult example of an *ars moriendi*, and the most edifying."[13] As Puppi has described, in the days leading up to the event, printed manifests blanketed the city's walls, announcing the hour and place of the execution. The state would typically effect absolute justice in the central Piazza Maggiore, where citizens would gather on the appointed day at the summons of a cacophony of church bells to witness the torment and death of the villain and, it was hoped, thereby quell any wickedness lurking in their own hearts (fig. 11). The mob, often festooned in carnival masks and electrified by anxious expectancy, would pack into the grand piazza and jostle for position. Clad in their white cloaks prominently marked with a black

FIGURE 11. Pisanello, *Saint George and the Princess of Trebizond,* Verona (1438); detail.

skull and cross, the religious Morte Confraternity led the condemned through the city streets to the site of the execution.[14] Notably, Archbishop Prospero Lambertini was himself a member of the confraternity and, despite his high position, conducted criminals to the top of the scaffold, prayed with them for God's mercy, and remained as a comfort until their death. Even when capital criminals inconvenienced the state by avoiding prosecution during Carnival, the unclaimed corpses of the indigent that would then serve for the Public Anatomy were seen to exude a similar odor of iniquity as those whose crimes officially warranted the final postmortem humiliation of dissection.

An affiliated protocol and ceremonial rules also governed the discharge of the Public Anatomy, with their procedures set in motion several weeks before the actual dissection of the criminal's corpse. According to eighteenth-century rules governing the "Ceremony of the Anatomy Lesson,"

> In the first days of January, the University Prior customarily contacts the Anato-mist at which time the Anatomist as a rule asks him when he would command the commencement of the Anatomy. With decorum they then reach a cordial agreement on the day. The day set . . . the most illustrious and excellent Deacon informs the most illustrious and excellent Administrators of the University such that the Anatomist and Surgeon are called to the Public Palace and are assigned the [Public Anatomy]. . . . The Anatomist then invites the most eminent Cardi-nal Legate to the first lesson . . . [along with] the Vice Legate . . . the Gonfalo-niere and . . . the Prior of the Anziani . . . the University Prior . . . and the most illustrious and reverend Archdeacon. Finally, it is customary to invite the most eminent Signor Cardinal Archbishop.[15]

Indeed, Archbishop Lambertini was a regular witness to the opening lesson that occurred, as Giovanna Ferrari has brilliantly described, in the smoke and shim-mer of candlelight reflecting off iridescent satin panels that swathed the walls of the second-floor anatomy theater in the Archiginnasio. With him, robed civic and religious authorities and university professors took their assigned seats. The prior, cardinal legate, vice legate, archbishop, and professor-anato-mist each had his own carved armchair (fig. 12). Musical accompaniment set a somber mood but offered slight distraction from the bone-penetrating chill in the room. Once the high-ranking officials were installed, university students quickly clambered into the available seating scattered among the three rows of tiered benches that ascended all four walls. Finally, space allowing, common citizens and foreign visitors could purchase a ticket for the remaining seats or for standing room in the back. From behind their carnival masks, these often unruly and vociferous spectators dared prolonged looks at the cadaver laid out on the marble dissection table. Its parts, now bleached and glistening, now wrapped in gloom from the oscillating torchlight at the head and foot of

FIGURE 12. Archiginnasio Anatomy Theater. Courtesy of Museo di Palazzo Poggi, Università di Bologna.

the table, were protected from the hands of the overly curious by a wooden balustrade.[16]

Each of the fourteen days of the Public Anatomy offered its own publicized disputation in Latin. The spectacle was, therefore, multiform. The carving of the cadaver by prosectors and surgeons that at once incited intellectual curiosity, fear, and rowdy taunts from the Carnival masques competed with the often near-riotous scholarly quarrel. For almost half the eighteenth century, a third spectacle was an added draw for visitors to Bologna's anatomy theater. From 1734 until the year of her death in 1778, Laura Bassi was obligated to showcase her uncommon learning as a regular disputant at the Public Anatomy. The refined female figure in hoopskirt and fur stole struck an artful contrast to the drama of the naked and dismembered cadaver at her side.[17] Her role in the dissection scene will be discussed at greater length later in this chapter.

A number of historians have elucidated the symbolic function of the Public Anatomy as social and sacred ritual.[18] Andrew Cunningham has sardonically likened it to a foxhunt, because essentially "it is about blood, death, and pleasure,"[19] the blood and death of the sinner-criminal and the ritualized pleasure of witnessing his public execution and dissection, which restored the body politic to its very marrow. Far apart from the needs of medicine and anatomy students, which were, in fact, little served by the public dissection, it at once reinforced traditional social hierarchies and belief systems and, as important, excited intense foreign interest, which in turn fueled municipal pride. According to Bolognese civic leaders themselves, the Public Anatomy satisfied "the interests of the splendor, the decoration and the honorific needs of the University and the whole city."[20] Copious municipal documents recount the vigorous mid-century drive to restore the now-diminished Public Anatomy to its past glory, as "singular and unique in Italy, if not the whole world."[21] A distinctly subordinate concern among civic leaders was the instruction of medical students, who had far better venues for their training in human dissection, namely the anatomical and surgical laboratories of designated hospitals and the household studios of their professors, where yearlong courses in practical anatomy were held. Spectacle thus clearly subsumed science in this dramatic exhibition.

The art historian William Heckscher has argued compellingly that the Public Anatomy should be read as the central act of a three-act morality play coproduced by church and state. Fixed characters, from criminal to executioner, cardinal legate to anatomist, acted out a set plot before a designated public of the powerful and, at times, popular classes. Performed in the first act was the capital execution of a criminal in the public square. The public dissection that comprised the second act meted out a *contrappasso* punishment against the body of the criminal, who had violated the body politic. The celebratory banquet of surgeons and civic leaders at the fulfillment of the penalty and the Christian burial of the criminal marked the restoration of social order and constituted the drama's denouement.[22] It was theater, however, with potent mystical consequences.

Through the ritualized public execution of the criminal body and its desecration by means of dissection, a conversion—indeed, a transubstantiation—potentially took place of a doomed into a saved soul, a sinner into a saint.[23] In his exquisite documentation and analysis of the case of the young Bolognese maidservant Lucia Cremonini, who was hanged in 1710 in the Piazza Maggiore for the crime of infanticide, Adriano Prosperi gives breath and pulse to the theory that the capital execution could, in the ideal, become a martyrdom reconciling both the criminal and the witnesses to her death into God's grace. Prosperi writes:

An extraordinary metamorphosis: Lucia, with eyes fixed on the votive tablet
(that is, on the image of the crucified Christ held by her Consoler), making her
way to the gallows to offer there an example of Christian death, is no longer the
condemned and abhorred infanticide. She is a courageous woman who makes
a display of her repentant self and accepts her own death. She does so in such a
deeply convincing way as to appear in the eyes of her brother Consolers a saint
going to martyrdom, joined to the community of believers by common prayers
and ready to face her journey toward Paradise.[24]

The dissection of the criminal body that followed a capital hanging (though
not, it seems, in Lucia's case) extended the punishment beyond death and, it
was hoped, the possibility for grace among the contrite.[25] As Katharine Park
has observed, "like the body of Christ, who died like a criminal, mutilated
on the cross, or like the scattered bones of long-dead martyrs," the broken
and offended corpse of the anatomized criminal was a potential "conduit for
divine grace."[26]

The Public Anatomy therefore served the twin functions, both fundamentally symbolic, of disciplining nature and fallen human nature. It was a
conquest of the secrets of the body through dissection, and of the sinful soul
through the redemptive retribution of suffering, death, and the defilement of
the sinner's bodily remains. The Anatomy Museum that Pope Benedict XIV
would establish would artfully integrate, indeed virtually incorporate in colored wax, the spectacle, the symbolism, and the science of Bologna's Carnival
Dissection.

The Pope's Modeler

When Archbishop Lambertini encountered the anatomical waxworks of
Ercole Lelli in 1731 while touring the refurbished Natural History Room of
the Institute of Sciences, he instantly recognized a potent new instrument for
reinvigorating Bologna's distinction in the realm of anatomical science. On
display were Lelli's vivid facsimiles of two human kidneys: the first normal,
the other an anomalous renal fusion known as a horseshoe kidney, modeled
directly on the "monstrous" organ that the prosector Lorenzo Bonazzoli had
discovered during dissection (fig. 13).[27] The archbishop grasped the didactic
promise of wax anatomical models to exhibit authentic and lasting views of
the anatomized body that in dissection would begin to molder on the first
cut. Even more important, he understood the power of Lelli's waxworks to
dramatize, for an international audience in the astonishing and uncanny idiom
of impressionable, colored wax, Bologna's return to preeminence in the medical sciences.

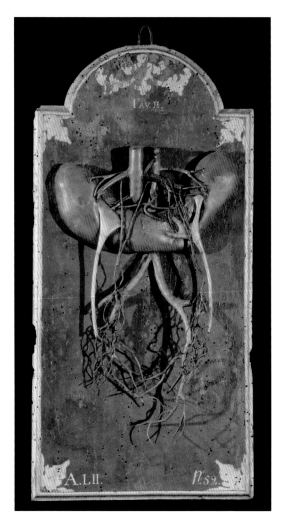

FIGURE 13. Ercole Lelli, horseshoe kidney, wax. Courtesy of Museo di Palazzo Poggi, Università di Bologna.

Lambertini at once conceived an anatomy museum exhibiting Lelli's life-like waxworks to be installed in the institute. In a letter of 7 February 1732 to the institute's Assunteria, or Governing Body, the archbishop declares his desire "to open an Anatomy Museum of the Institute in which the Artist Lelli would work, creating in wax five life-size figures all of whose parts would be configured in wax."[28] Five days later, the artist submitted a proposal and budget for the production of the five écorchés demonstrating the complete myology and osteology, or muscle and bone structure, of the human body, to be completed in four years.[29]

The pope's chosen modeler Ercole Lelli proved an ideal partner for dramatically staging Bologna's resurgence in anatomy. His focused stage management of his own achievements in that city's social and cultural arenas clearly enhanced this partnership.[30] Under the influential tutelage of Giampietro Zanotti, the secretary of Bologna's Clementina Academy of Art within the Institute of Sciences, Lelli excelled as a sculptor.[31] But rather than take his place among the herd of young and established artists replicating conventional sculptural styles and motifs, he strategically set himself apart in this historic center of anatomical studies by developing expertise in anatomical sculpture and design.[32] As Lelli's teacher, Zanotti originally encouraged him to perfect these skills by observing dissections at the hospital and mortuary of Santa Maria della Morte. Zanotti in fact forcefully defended this specialized and exceptional training against strident criticism from within the Academy of Art. However, he disapproved of an overweening pursuit of anatomical knowledge, which he always considered a means to proficiency in the art of representing the human form, but not an end in itself. He frowned on Lelli's eventual performance of his own dissections at the hospital as well as his experiments with wax injections of veins, capillaries, and other body parts. As is plain from an extended advice-poem written by Zanotti to Lelli, the teacher sought to discourage these practices and to emphasize the proper auxiliary function of anatomical studies in the artistic preparation of his student:

Dimmi, Ercolin, che fai,	Tell me, Ercolin, what are you doing,
che più non veggioti	that I haven't seen you
Da lungo tempo in qua? Dov'hai tu l'animo	Here for so long? Where is your mind
Ora rivolto, e in qual parte lo studio	Now turned, and to which part the study
Della nostr'arte? Io credo, che non scortichi	Of our art? I trust you no longer flay
Più, per apprender notomia, cadaveri,	Cadavers to learn anatomy,
Che ne sai quanto a dipintor richiedesi	For you know that to paint requires
Cento altre cose sonvi, che abbisognano	A hundred necessary other things
Senza le quali notomia non giovaci.	Without which anatomy is useless.[33]

Lelli disregarded his teacher's advice, however. He had already carved his first anatomical figure by the age of twenty-six, an exquisite miniature écorché avidly copied by other young artists.[34] He of course modeled on actual dissected organs the normal and malformed kidneys that had so impressed the archbishop. Near the time of Lambertini's viewing of these wax kidneys, Lelli actually was attempting to further solidify his position as Bologna's chief anatomical sculptor by inserting himself into the Senate's project to restore the Archiginnasio Anatomy Theater. Determined to join his name and fame to this cultural monument of old and *new* Bologna, he offered to replace at his own expense the grandiose caryatids upholding the lector's throne.[35] Lelli carved from sturdy linden wood two life-size male figures demonstrating the surface musculature, one from the anterior and the other from the posterior view (fig. 14).

FIGURE 14. Ercole Lelli, anatomical écorchés, wood. Archiginnasio Theater. Courtesy of Museo di Palazzo Poggi, Università di Bologna.

His contemporary biographer Giovanni Fantuzzi (1718–1799) recorded that the artist used more than fifty cadavers as the empirical basis for his two figures, and constructed preliminary models for his muscle men from actual skeletons. Notably, he broke down the numerous intact skeletons and reconstructed from this vast collection of disconnected bones more perfectly proportioned figures as required by Vasari. Lelli would later write in his *Anatomical Compendium* that his overarching aim was to "express with elegance and propriety the human figure" for the benefit of artists. His wooden muscle men were modeled on these ideal skeletons, which he erected and overlaid with cloth and wax carved in the exact likeness of the musculature. Lelli clearly grasped the emblematic significance of the theater, and by literally carving his spectacular and aesthetically transformative mark on it, he tied his own cultural status to that of this illustrious place.[36]

Lelli's personal ambitions in originating a new art of anatomy thus perfectly complemented the pope's civic ambitions to restore Bologna's past distinction in the medical sciences by establishing the Institute of Sciences' Anatomy Museum. Glory for each man hinged on his signature renovation of the Bolognese cultural landscape, and the Anatomy Museum was a test and testament of their power to re-form La Dotta. As we have seen, both men ultimately looked to the annual dissection scene within the Archiginnasio Theater as a template for the ideal synthesis of spectacle and science that the Anatomy Museum would place on permanent display.

Competing Ideals

Continual problems with local funding obstructed Lelli's progress on the museum project, however, and regular work would not begin until 1740, when Lambertini became pope and could then fund the immense project himself. It is clear from his correspondence that Lambertini's passion for the museum only heightened during the ten-year delay. Indeed, although Pope Benedict XIV Lambertini never returned to Bologna during his eighteen-year pontificate, he continued to micromanage from Rome, at times obsessively so, key aspects of the cultural life of his city. It would appear that he saw his legacy defined nearly as much by his success in elevating Bologna's international cultural status as by his impact on greater Christendom. His unflagging determination to erect the wax anatomy museum reveals the competing objectives that marked his efforts to restore Bologna's cultural preeminence. Plainly evident in his commission of the museum is tension between his patronage of the *new* science, its empirical methods and modern machines, and his more provincial ambitions to expand by means of scientific spectacle the international celebrity of his native city.[37] Without doubt, Lambertini sought to modernize the practice of science in Bologna. His own intellectual formation and vocation were steeped in the literature of the new empirical science and philosophy, interests that became an important thread running throughout his public and private writings. In his most influential religious work, *De servorum dei beatificatione et beatorum canonizatione* (On the Beatification and Canonization of Saints), church doctrine and science can in fact be seen to collaborate in direct and unprecedented ways to expose false from true claims to saintliness. In the fourth and last volume of this major work, perhaps not surprisingly, Lambertini inserts a brief, approving history of Bologna's Institute of Sciences in which he alludes to the influence its academicians had on his own scientific formation.[38] Yet notwithstanding his authentic advocacy for science in the substantive form of policies, funding, and instruments to facilitate the work of the institute, his simultaneous and, at times, overriding concern was the international renown of Bologna itself. Lambertini's sponsorship of Laura Bassi is a relevant case in point. An axis figure in the pope's most famous reform initiatives, Bassi epitomized his patronage of the *new* science and learned women; however, she was also a foremost means in his campaign to upgrade and showcase Bologna's cultural distinction. All these roles can be seen to converge within the context of the Archiginnasio Anatomy Theater.

Bologna Minerva

Leonardo Sconzani's vibrant watercolor miniature of Laura Bassi's 1732 thesis defense shows the twenty-one-year-old philosophy student orating from the lectern in the Sala del Consiglio of the Palazzo Pubblico, seated ramrod straight with her arms raised, expressively gesticulating toward the red-canopied throne where Cardinal Archbishop Prospero Lambertini sits alongside the cardinal legate. The room swells with an exuberant crowd of elected officials in powdered wigs, high-ranking clergy, university professors, and select aristocratic men and women (fig. 15).[39] The internal frame in which the scene is set is a coat of arms decorated along its periphery by nine crests of Bolognese municipal and religious leaders.[40] The center is open, however, and serves as both a window onto the historic scene and a new heraldic seal of the city for which Laura Bassi, the first woman to receive a degree from the University of Bologna, is the central blazon. As manifest in this image and each of the watercolors in the government's collection lauding Bassi's civic honors, Bologna's modern Minerva uniquely embodied the city's Enlightenment progress.[41] Findlen reasons that "publicizing Bassi's accomplishments, and enhancing them beyond anything achieved by earlier learned women, would add luster to the reputation of Bologna."[42] Prospero Lambertini was instrumental in coordinating and amplifying Bassi's publicity and thereby that of the city that conceived her.

From his first year as archbishop in 1731, Lambertini assisted Bassi's rise. His presence at her thesis defense, at the conferral of university laurels on her, and at her subsequent public lectures in the Archiginnasio during the Carnival Dissection patently blessed the granting of institutional authority to this learned woman and established a new precedent and icon for Bolognese cultural distinction. Indeed, as Cavazza painstakingly documents, the archbishop had a hand in nearly every detail of the defense ceremony, including Bassi's

FIGURE 15. Leonardo Sconzani, Laura Bassi, 1732 thesis defense. Anziani consoli, *Insignia degli Anziani*, vol. 13, carta 94a. Courtesy of the Bologna State Archive. Used by permission of Ministero per i Beni e le Attività Culturali.

transport to the Public Palace in the carriage of the Gonfaloniere, or chief magistrate; her noble female escorts; the lavish ornamentation of the Sala del Consiglio; and indeed the decision that the ceremony be held at the Palazzo Pubblico instead of a church, as was customary.[43] Lambertini's advocacy led to Bassi's well-paid and vaunted appointment by the Senate as university lecturer in universal philosophy nine months after she took her degree, and in 1745 to her honorary membership in the institute's Academy of Science, renamed the Accademia Benedettina for Pope Benedict XIV himself.[44]

Much as Lambertini strove to renew his native city's legacy in the medical sciences, so too he sought to revitalize Bologna's standing as a historic wellspring for exceptional learned women. A standard catalogue of illustrious Bolognese women from the Middle Ages and the Renaissance provided the legitimizing precedent for his spectacular public promotion of Laura Bassi and other contemporary *scienziate*.[45] Beyond his support of Bassi, Lambertini advocated the conferral of a university degree on the Rovigan physicist Cristina Roccati in 1751.[46] One year earlier, he appointed to a university lectureship the Milanese mathematician Maria Gaetana Agnesi, whose two-volume book *Istituzioni analitiche ad uso della gioventù italiana*, published in 1748, was deemed by the French Académie des Sciences the most important work on differential and integral calculus of the day. Although she did not come to Bologna to fulfill her appointments, Agnesi, like Bassi, was also installed as an honorary Benedettina, a member of the institute's preeminent Academy of Science.[47]

As will be seen, Benedict XIV also became a protector of Anna Morandi at a precarious moment in her anatomical career, just after the death of her husband, when her family's financial distress heightened the possibility that another European city might lure her away. Thus was Pope Benedict XIV, as Cavazza has somewhat romantically described, "the deus ex machina who . . . ensured that Bologna, the second city of the Holy See, became during the Settecento the new 'paradise of women.'"[48] Yet the exceptional institutional authority granted learned women during the Italian eighteenth century, authority unmatched anywhere else in Europe, was not confined to Bologna, but extended in varying degrees across the peninsula, where it was touted by civic leaders as a sign of their states' Enlightenment social and scientific progress.

Lambertini's benefaction of learned women no doubt issued from genuine admiration as well as a desire to challenge certain conventional beliefs in the essential inferiority of the female sex so prolifically debated from Venice to Naples at the time.[49] Yet the public promotion of women scientists, poets, and artists derived equally from his desire to captivate the Enlightenment Republic of Letters. And on this front, Lambertini was quite successful. The noted literary republicans Voltaire, Lalande, Burney, and Johann Jacob von Volkmann, and the American physician John Morgan, among many others, wrote of their admiration for Bologna's woman philosopher and for the city that aided her ascent.[50] Bassi, for her part, fulfilled her formal duties as exceptional civic attraction and hostess by meeting foreign dignitaries at official assemblies and private *conversazioni*. Burney's comments typify Grand Tourists' impressions of these encounters:

> In Italy, degrees are given to scientific Ladies; and at this time one of the most distinguished is la Signora Laura Bassi of Bologna. . . . This lady is between fifty and sixty; but though learned, and a genius, not at all masculine or assuming. She seems perfectly acquainted with the merit of all the learned and scientific men in Europe, and was particularly civil to the English in encomiums on Newton, Flamsteed, Hailey, Bradley Franklin, and others. This Lady has made an uncommon progress in Electricity, has invented new machinery, made new experiments, and fairly earned her title of *Dottoressa*.[51]

Most indicative of her role as municipal marvel were, however, the aforementioned command-performance disputations she gave at the Public Anatomy.

As the official watercolor insignia of the event illustrates, with her hoop-skirted female attendants at her back in conspicuous defense of her honor, Bassi took up her appointed role as Bologna Minerva, delivering her first Public Anatomy lecture in 1734 with an ideal mix of humility and virtuosity to the city's foremost anatomy professor, Domenico Maria Gusmano Galeazzi, seated on the lector's throne, while the prosector Lorenzo Bonazzoli dissected the male corpse (fig. 16).[52] The two most potent emblems of the *new* Bologna, the anatomized human body and the woman philosopher, thus occupied the same stage in this scene, reenacted predictably year after year until Bassi's death in 1778.

Laura Bassi and Anna Morandi were thus linked by their intimate experience with the anatomized body. The connotations of a woman hard by a dead and dismembered (male) body are vast and traverse the symbolic terrain of such distant feminine paragons as the Madonna of the Pietà and the petrifying Medusa.[53] These and all between are certainly at play in the Carnival Dissection and the household anatomy demonstrations in which Bassi and Morandi respectively engaged. However, as this study will illustrate over the course of subsequent chapters that document Anna Morandi's lifework, she presented a striking contrast to her celebrated contemporary. Morandi's entrance onto the Bolognese cultural stage was not secured by her attainment of a university degree. She was never asked to give a public oration and enjoyed no escort in the Gonfaloniere's gala carriage. Volumes of occasional poems were not written in her honor. Most important, she assumed a different public role from Laura Bassi's in the anatomical dissection scenes she herself directed and regularly performed in her home studio for medical practitioners and foreign visitors. Of a decidedly lower cultural class with her practical, hands-on knowledge of the anatomized body, Morandi fit the profile of neither the conventional illustrious woman (*donna illustre*) nor the state Minerva, Athena, Venus, etc., touted in Bologna and other aspiring Enlightenment city-states across the peninsula as emblems of a "newer" science and vanguard Republic of Letters.

Paradoxically, however, Morandi and Bassi forged another implicit bond through their defiance of the limits of their civic roles. Within their home studios, each woman defined her role and work in a way she could not on the outside. Morandi cultivated prominent patrons across Europe and drew a distinguished specialized public to her laboratory for the expert work in anatomical research and design she conducted together with her husband and then on her own after his death. It was here that she defined herself as Lady

FIGURE 16. Leonardo Sconzani, Laura Bassi's 1734 anatomy oration. Anziani consoli, *Insignia degli Anziani,* vol. 13, carta 105a. Courtesy of Bologna State Archive. Used by permission of Ministero per i Beni e le Attività Culturali.

Anatomist against the official designations of artisan, modeler, and anatomical demonstrator. As municipal documents make clear, Bassi was prohibited "by reason of her sex to teach at the University" (ratione sexus publico in Archigymnasio doceat),[54] unless expressly called on by the Bolognese Senate, as in the case of the annual Public Anatomy. She was therefore unable to perform, except on specified and rare occasions, the duties conferred on her as university lecturer. Indeed, only in the last two years of her life, 1776–78, did she finally become a regular (as opposed to honorary) professor of experimental physics at the Institute of Sciences, never the university. Beginning in 1749, with support from her husband and colleague, the physician Giuseppe Verati, Bassi taught in her home the cutting-edge science she could not practice in public. And students and visitors of high social and academic rank clambered to become her pupils there.[55]

Each woman thus sought to establish in her home laboratory a certain level of autonomy from the momentous enterprise of the *new* Bologna in which she held an assigned, albeit distinct, part. Nonetheless, the sweeping project of the city's cultural renewal infiltrated in key ways their personal spaces and aims. Superintended by Lambertini almost certainly with even more fervor after he took up residence in Rome in 1740, Bologna's Enlightenment renaissance converged at select cultural sights and human attractions. Bassi and to a lesser degree Morandi were two of these. The new pope, however, directed his attention most earnestly to showcasing Bologna's ascent by raising his wax Anatomy Museum.

The Papal Commission

From his papal throne, on 1 December 1742, Lambertini commissioned Ercole Lelli to sculpt in wax and according to nature eight life-size figures of the complete muscular and skeletal structure of the human body for the founding of Italy's first Museum in Italy of Human Anatomy or Internal and External Anthropometry, as it was called.[56] The written commission directed the "excellent sculptor" to begin the series with a male nude that would serve to illustrate "not only the symmetry but . . . the diverse sites and regions of the human body," and to follow this with its counterpart, "a nude woman," demonstrating "the diversity of her sites and her parts."[57] The last two figures in the series, representing the skeletal structure, would follow four myological écorchés. They too would "demonstrate the difference that passes between the two sexes, even in the figure and disposition of various bones."[58]

Sexual difference was therefore a defining theme of the Anatomy Museum. Little has been made of this fact in relevant historical accounts, however.[59] That the pope sanctioned the side-by-side representation of anatomized male and female bodies for the purpose of their systematic comparison conformed to an ascendant trend in European anatomical science and design. Pope Benedict XIV was, of course, keen to sponsor scientific work at pace with or ahead of the latest developments across Europe. Methodical analysis of the anatomical and physiological differences of women formed the basis of the *new Science of Woman* that proliferated during the eighteenth century as a principal subset of the *Science of Man*, which had taken root and flourished from the late seventeenth century.[60] Questions regarding women's essential physical composition in turn underpinned the vigorous public debate waged across the Italian peninsula about the place and purpose of women in society and the state.[61] As Londa Schiebinger has shown, the eighteenth century saw sharply increased attention not only to the inherent differences of the male and female reproductive systems and genitalia, but to all the parts and substances of the body, including the deep skeletal structure.[62] The pope's directive to juxtapose wax anatomical Eves to anatomized Adams indicates a calculated investment in the period's campaign to define and taxonomize sexual difference, to the very bones. Attention to the anatomical differences between the sexes promised to further stimulate international interest in the Anatomy Museum and the work of the Institute of Sciences. Just as Bassi's extraordinary feminine presence at the Carnival Dissection increased its fascination for an international public, a comparative focus on the female anatomy would undoubtedly draw more attention to the museum.

As will be seen in chapter 6, Anna Morandi, too, focused extensively in her notes and waxworks on the structure and functions of the reproductive body.

Strikingly, however, Morandi's analysis centered on the male body, making this a leitmotif of her oeuvre. Moreover, in contrast to contemporary trends in the science and the specific juxtaposition of the sexes put on view in the pope's museum, she refrained throughout her oeuvre from direct comparisons of male and female bodies. Her eschewal in her analysis of the ascendant question of sexual difference will be discussed at length in chapter 5.

Art over Science

In his project proposal, Lelli asserts the self-aggrandizing claim that his anatomical waxworks "would be of the greatest utility and importance for surgery."[63] Yet all traces of the use of the models for the purpose of practical medical training are absent in the papal commission. The pope was unequivocal that Lelli's wax anatomies were to be considered museum pieces and models for artists to train from, not instruments for teaching medical students. Authentic medical training, Lambertini insists, required the dissection of cadavers by "Men expert in Anatomy and the most renowned Surgeons with suitable instruments in the designated laboratories of the public hospitals."[64] He thus drew a clear distinction between the practice of anatomical science and its artful exhibition. Midway through Lelli's project, in 1747, the pope issued a *motu proprio* to specify further the pedagogical and civic functions of the Anatomy Museum. His new statement makes explicit the previously tacit connection between the annual spectacle of the Public Anatomy and the perennial anatomical wax museum:

> Given the critical place in which the People hold the Public Anatomy, which is undertaken every year by a Medical Lector in a very distinguished fashion, superior to that performed at any other University, and that for the greater benefit of novices there is always an adept dissector who, once the Anatomist's lesson and disputation are finished . . . assists the Anatomist in the demonstration of the parts he prepared . . . since there is nothing comparable at the Institute . . . we recognized the deficiency in this area. . . . We therefore resolved to have made, at no small expense, by Ercole Lelli, expert not only in Painting and Sculpture, but by his extensive study, very well-versed in Anatomy, eight life-size wax Statues demonstrating distinctly the muscle and bone structures, which in the traditional Public Anatomy are treated generally. As a consequence [of such treatment, these structures] cannot be adequately seen, nor can they sufficiently convey to Painters and Sculptors . . . those principles needed by Beginners in these Arts and Profession.[65]

The wax museum would thus serve to replicate and indeed enhance the most constructive features of Bologna University's famous Public Anatomy. It would offer unhindered views of the anatomized body to a diverse public.

While the models would not serve the specialized aims of medical science, Lambertini underscored their use as elemental teaching tools, especially in the remedial instruction of neophytes in anatomy, specifically art students. At least as important as their demonstration of the locomotor apparatus of the body—its bones and muscles—for figure drawing and sculpture, Lelli's waxworks would, moreover, add to the luster of the institute just as the Public Anatomy fueled the fame of the university. Like the Carnival Dissection, the museum was thus a means for advancing Bologna's celebrity as a center of science, though not necessarily practical anatomy itself.

In at least one crucial way, however, these two forms of Public Anatomy sharply diverged. The cultural and civil framework of the museum obliged a very different demeanor of its spectators than that of the Carnival Dissection with its ostentatious civic, academic, and religious leaders, unruly masques, and bickering scholars. A site of Enlightenment inquiry, the Institute of Science's Anatomy Museum required a subdued and civil witness to these bodies unmade. The models would awe rather than inflame viewers through their dramatic illustration of the ideal taxonomy and the harmony and mechanics of male and female bodies, as well as through their moralization of human life matter.

Tableau Mort

In fulfillment of the terms of the papal commission, Ercole Lelli's anatomical series illustrated incrementally, from shallow to deep, the body's ideal musculature and bone structure as well as the essential proportional differences between the sexes. His figures were actually a composite of the perfect bony parts of scores of male and female cadavers that he had dissected, painstakingly reassembled, and covered in wax.

Lelli explicitly followed the dictates of the canon of anatomical design, maintaining the same order of body parts illustrated in the atlases of "Vesalius, Valverde and Eustachio."[66] Vesalius's *Fabrica* was, in fact, a key exemplar of anatomical illustration for his waxworks. It is perhaps unremarkable, then, that the anatomized body proved an insufficient subject for Lelli, one that he felt obliged to sublimate in terms of both its physical and its metaphysical significance. An embedded, synchronous narrative of the demise of the soul underpinned his ideal rendering of the body's internal framework.

Adam and Eve lead the ranks of Lelli's anatomical figures (figs. 17–18). The whole, naturalistically colored male and female nudes placed in classical contrapposto epitomize the neoclassical aesthetic ideal, evincing their disgrace through archetypical poses and suppliant facial expressions.[67] While Lelli's fallen couple expresses subtle movements indicative of their internal

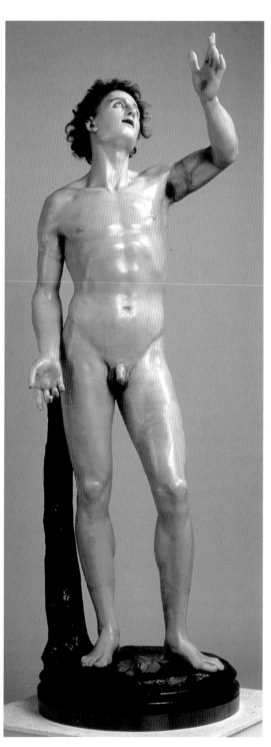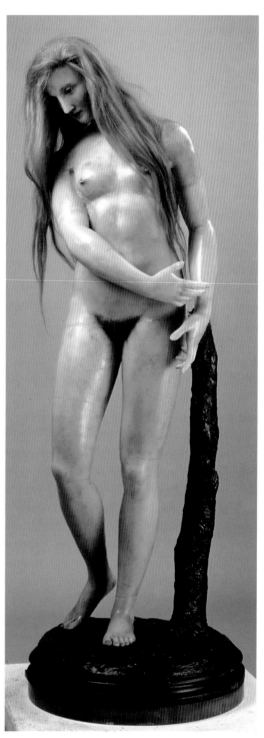

FIGURE 17-18. Ercole Lelli, Adam and Eve, wax, wood, bone. Courtesy of Museo di Palazzo Poggi, Università di Bologna.

turmoil, he poses them in ways to manifest a range of both muscular tension and emotion. With head bent and eyes downcast over her right shoulder in an attitude of shame and remorse, Eve twists her torso to the right while her right arm extends across her trunk to the left. Her waist-length, flaxen hair falls over her shoulders and breasts, paradoxically accentuating her nakedness by the insufficiency of the veil it provides. The display of pubic hair further embellishes her nudity. By contrast, the curly black-haired Adam, as so many Saint Sebastians, exhibits his nakedness in an unabashed full-frontal posture, his muscular left arm raised in beseeching supplication. Adam and Eve provide a striking contrast, physical as well as metaphorical, to each other as well as to the progressively dissected figures that follow.

Next to Adam and Eve stand two sets of écorchés, or flayed men (figs. 19–22), that represent in stages the surface to deep musculature. Like the nudes, the *scorticati* simultaneously serve scientific, aesthetic, and moralizing ends. Posed to emphasize distinct muscle groups and bones, this tragic, extra-naked cast of the fallen also embodies its state of disgrace by means of facial expressions, posture, and gestures of abject fear and despair. While illustrating the head-to-toe, superficial skeletal muscles and ivory tendons that position and move the body, as well as the lymph node pockets at groin and armpit, the first écorché stares, wide-eyed and grimacing in terror, from behind his defensively raised left arm. Further down the line, a more pitiable écorché, whose skeletal right hand rests on his heart—a stance that seems to incarnate the poet William Cowper's "Hollow-eyed Abstinence, and lean Despair"—displays the deepest muscular attachments to the skeletal structure.[68] His despondent, level gaze is made more striking by the blood-red orbicularis oculi muscles that frame his gray-green eyes.

Lelli concludes his *tableau mort* with two skeletons formed from the actual bones of scores of cadavers (figs. 23–24). This pair serves as a congruent match-set to the intact male and female nudes with which the anatomical series began. In his demonstration of the male osteology, Lelli follows Bernhard Albinus's definitive portrait of the male skeletal structure that had dominated anatomical illustration since its appearance in 1734. However, he supplements Albinus's male figure with a complementary, idealized female skeleton. Lelli was, like Albinus, chiefly concerned with the representation of universal anatomical archetypes.[69] His paradigmatic male skeleton thus boasts a hefty skull; a strikingly wide rib cage; strong; squared hips; a thick spinal column; and long arms and hands that powerfully grasp a heavy scythe. By contrast, the female skeleton has a slender frame; a compact rib cage; a delicate neck and spine; a small though appropriately wide pelvis; slight legs and feet, the right raised somewhat whimsically onto the toes; and narrow arms and hands,

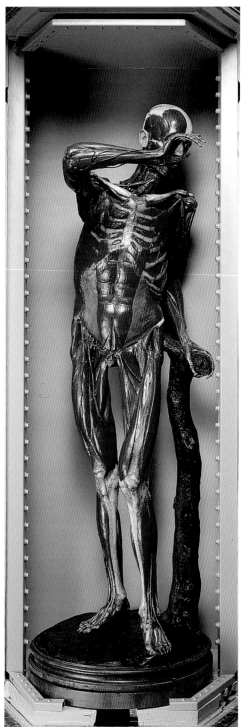

FIGURE 19–20. Ercole Lelli, figures of surface to deep musculature, wax and bone. Courtesy of Museo di Palazzo Poggi, Università di Bologna.

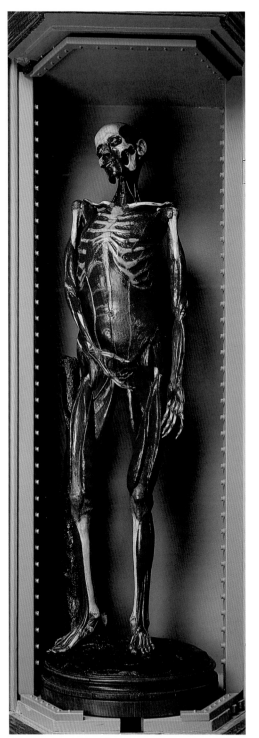
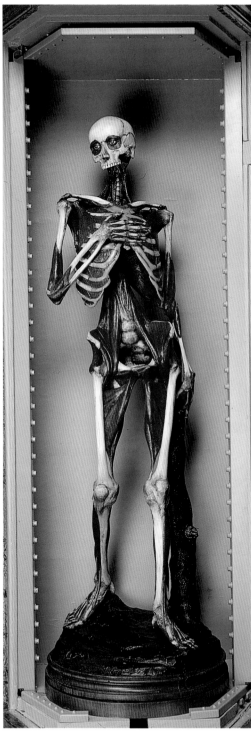

FIGURE 21–22. Ercole Lelli, figures of surface to deep musculature, wax and bone. Courtesy of Museo di Palazzo Poggi, Università di Bologna.

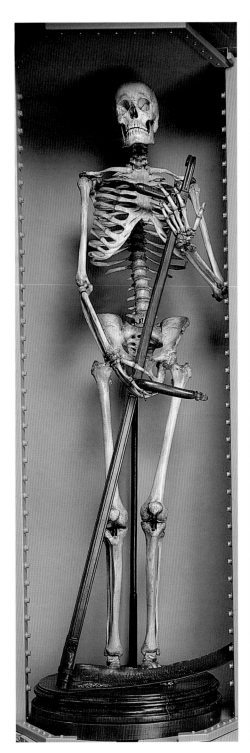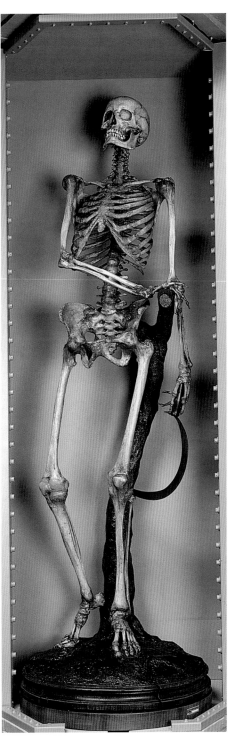

FIGURE 23–24. Ercole Lelli, male and female skeletons. Courtesy of Museo di Palazzo Poggi, Università di Bologna.

the left of which rests lightly on her scythe. Lelli's is a conventional dualistic conception of the sexes—man: large and powerful, woman: small, agile, and reproductively endowed—rooted in traditional anatomical study as well as in an Adamic taxonomy that the series explicitly evokes.

Beyond their scientific and taxonomic functions, like the figures of Adam and Eve, the two skeletons contravene the boundaries between anatomical science and memento mori. Brandishing immense iron scythes, the figures are obvious emblems of Grim Death and the retributive justice, both earthly and divine, at play in the Carnival Dissection scene. The final panel in the wax triptych that began with the First Couple in the implied Garden, followed by the mournful company of Adam's errant progeny, the paired skeletons serve as an unequivocal moral epitaph to the story of man's original Fall and inevitable faltering ever after.

Ideal visitors to Lelli's allegorical écorchés were not, therefore, the ambitious disciples of Valsalva, Malpighi, and Morgagni, doctors and students of medicine, but rather an assortment of elite and middling, native and foreign spectators similar to those attendant at the Public Anatomy. They were summoned to look on the perpetual dissection of these archetypal sinners with the eyes of righteous judges as well as mortified sinners. Lelli's idealized vision of the body thus predicated the truth-value of anatomical sculpture not merely on its exacting replication of nature, but also on its classical aesthetic and moral effect.

The New Reliquary

Visitors to the museum entered by way of the Room of Light, the last of three laboratories devoted to physics. As they crossed the threshold of the museum, they encountered eight glassed armoires lining three of the four walls and rising nearly to the ceiling (figs. 25–26). To fully view Lelli's wax effigies within, guests would have to walk the perimeter of the room and stand before each case, moving from the intact figures of Adam and Eve and the four progressively unmade muscle men to the two sexed skeletons that closed the series. The marble dissection table at the center of the room was the real and symbolic altar of these bodies' unmaking.

This temple of knowledge evoked a compressed church space in which symmetrical niches displayed looming sculptures of saints. The architecture and layout of the room, with its compartmentalized specimens systematically displayed against the walls, also recollected the seventeenth-century and more recent natural history cabinets for which Bologna, and indeed the Institute of Sciences, was famous. At the same time, the actual bones of myriad cadavers reassembled, posed, wrapped, and ribboned in wax and finally housed in

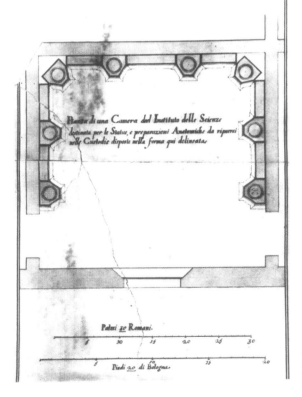

Pianta di una Camera del Instituto delle Scienze
destinata per le Statue, e preparazioni Anatomiche da riporsi
nelle Custodie disposte nella forma qui delineata.

Palmi 30 Romani.

5   10   15   20   25   30

5   10   15   20
Piedi 20 di Bologna.

FIGURE 25–26. Design of the
Anatomy Museum. Courtesy
of Museo di Palazzo Poggi,
Università di Bologna.

ornate faux-marble and glass armoires even more strongly suggested church reliquaries. Pilgrims to these crystalline shrines of the anatomized body would experience an alignment of facts and faith congruent with the Enlightenment pope's contemporaneous campaign against the popular cult of relics and *agnus dei*.[70] The museum thus offered a revised, Enlightenment cult of the science and symbolism of the anatomical body, a relic of philosophical truth and the surgeon's knife.

The "Lelli Problem"

Despite the spontaneous applause that erupted at their unveiling and the generous remuneration Lelli received for his work,[71] Lelli's wooden écorchés and subsequent wax anatomies set off a fierce polemic among Bolognese *accademici*, especially within the Clementina Art Academy. As Massimo Ferretti has shown in his recapitulation of the controversy, the argument centered on whether anatomical design was chiefly artistic or scientific in scope. At issue was the very purpose of anatomical design, or, more simply, whether artists should seek expert knowledge in anatomical science. The answer among many members of Bologna's Clementine Academy of Art was a resounding no.[72]

Propelling the controversy over anatomical wax design was a controlling and paradoxical belief that any artist who sought expertise in anatomy transgressed the bounds of his profession by both descending to the manual trade of the artisan and aspiring to become a scientist. As we have seen, Giampietro Zanotti, Lelli's teacher and the longtime secretary of the Clementina, disapproved of Lelli's specialization in anatomy and advised against the excessive training of artists in anatomy, which he believed ought to serve solely for the correct representation of the male nude.[73] Lelli's arch nemesis, the artist and biographer Luigi Crespi (discussed at greater length in chapter 3), gave vent to a far more contemptuous judgment in his *Lives of Bolognese Artists*. Lelli was the clear provocation for the scathing and protracted diatribe Crespi levels against those artists who sought distinction by their affectation of anatomical expertise:

> That doctors and surgeons speak of the thorax, the abdomen, the axillary fossa, the malleolus, the deltoide, the cuneiform cartilage etc. and that philosophists also speak of these is understandable. But that artists have a need to talk about the ilium, the pubis, the ischium, the acetabulum and about how many muscles move the eye, and how these muscles have their origin in the orbital fissure and their insertion in the corneal tunic and, not to tire the reader, that painters must talk anatomically about these internal parts of the human body is a grave fraud that in the end brings them nothing but a pact with fools and some self-serving ideas about their own merit. . . . Knowledge of the sciences will always be

praiseworthy and estimable, but to have the mastery of an art comprise those things that contribute nothing to the perfection of the art itself is idiocy and a sham. [It is] worse still if the best years of the poor, studious youth are employed in learning this affected and ridiculous string of names.[74]

Notably, Lelli's equally determined supporters extolled the artist as the supreme practitioner of a practical new art of the body's internal machinery. Among his most influential defenders were his former students Carlo Pisarri, Gian Lodovico Bianconi, and the philosopher and writer Francesco Algarotti (1712–1764), who stepped into the fray with his *Essay on Painting*.[75] Algarotti moves directly to the heart of the matter, arguing that "it is unnecessary for [the artist] to scrutinize the different systems of the nerves, blood vessels, bowels, and the like, parts which are far removed from the fight, and which may therefore be left to the surgeon and the physician."[76] Instead, the "young painter" must study the bones, muscles, and the connections between them in order to convey "their proper figure, situation, office and motion."[77] Ercole Lelli, Algarotti insists, went to "greater lengths . . . than any other master" to represent the differences, the color, and the configurations of the muscles for use by artists.[78]

Lelli thus had strong detractors of his anatomical art as well as commanding defenders, most notably the pope. However, the quarrel that erupted between Lelli and his chief assistant, Giovanni Manzolini, and spilled from Lelli's modeling studio into the streets and *piazze* of Bologna posed the gravest threat to his reputation as an artist. The crux of the "Lelli problem," as art historian Eugenio Riccomini has neatly designated it,[79] was widespread skepticism over Lelli's competence as an anatomical modeler and his claims to being the sole genius of the Anatomy Museum.[80]

The controversy peaked when Manzolini noisily quit the museum commission in 1747, accusing Lelli of stealing credit for his work. Unlike Lelli's two youthful and inexpert former assistants, who had also resigned from the prestigious commission, Manzolini was Lelli's professional peer (both were born in 1702) and, as Miriam Focaccia has documented, former classmate at the Clementina Academy of Art.[81] Manzolini possessed, moreover, training and professional qualifications comparable to those of the pope's modeler.[82] By the time he accepted the job of assistant, he had apprenticed in the studios of two major Bolognese artists, Giuseppe Pedretti and Francesco Monti, studied at the Academy of the Nude, and established himself as a local artist of certain fame. His enduring passion became human anatomy, which he studied on his own in the hospital of Santa Maria della Morte and as an apprentice in Lelli's workshop, developing uncommon skill in dissection and anatomical wax sculpture.[83] And while Lelli openly affirmed that his anatomical proficiency focused specifically on those parts of the body (muscle and bone) crucial for depicting the body's form and movement in art, Manzolini aspired to comprehensive, expert knowledge per se of all the body's anatomical components and systems.

Numerous extant accounts of the controversy leave little doubt that Lelli's role in the actual construction of the wax anatomical figures was markedly circumscribed, and that his assistants did the brunt of the work.[84] It is nevertheless clear that the magnificent and moralizing pageant of wax bodies created for the pope's museum embodied Lelli's artistic concept and neoclassical aesthetic. The two artists undoubtedly clashed, therefore, not merely over the distribution of labor on their race up the ladder of ambition, but also over their conflicting views of the correct object, method, and aesthetic of anatomical waxwork.

Intriguingly, Jacopo Bartolomeo Beccari (1682–1766), a leading light and the first Chair of Chemistry at the University of Bologna, wrote a furtive letter to Flaminio Scarselli (1705–1776), secretary to the Bolognese Embassy in Rome, saying that Lelli's "formidable jealousy" of Manzolini's talents had forced Manzolini from the project, while "pleasure" (*piacere*) above all else inspired Manzolini's devotion to anatomical science and his partnership in this study with his wife, Anna Morandi.[85] Immediately upon his departure from the museum project, the couple opened their home anatomy school and wax modeling studio. Beccari's letter, in fact, suggests that they had begun to work together in modeling anatomical wax figures even before Manzolini quit the commission.[86] Lelli's unrelenting envy would nevertheless hinder their local approval as well as lock them out of the Anatomy Museum of the Institute of Sciences that Lelli guarded as his own.

The force of his jealousy would, moreover, quickly shift to Anna Morandi as word spread of her rare talent as an anatomist and modeler. Four years before the Anatomy Museum opened, Lelli wrote an aggrieved and anxious letter to the pope's physician in Rome, Marc Antonio Laurenti, to learn if the pope had heard news of the Lady Anatomist. Laurenti reassured him, "With respect to your last, in which I hear that you had been told that the Pope was intrigued by news of a talented woman now making anatomical statues in wax, and in which you also informed me about what she is making and how, I can tell you that on the subject, the Pope never, and I mean never, said a word about her."[87]

Rival Collections

As Pope Benedict XIV had envisaged, upon its inauguration in 1752 the Anatomy Museum instantly enthralled Grand Tourists, who arrived from all corners of Europe and beyond to view the spectacular waxworks of Bologna's Institute of Sciences. Yet another collection of anatomical wax sculptures on display in the shadow of the pope's museum had been drawing a steady stream of visitors for several years, not only from among the distinguished foreign guests but also the local medical students and practitioners. They came to see the Lady Anatomist, Anna Morandi, give regular demonstrations of her and her husband's innumerable wax models of the component parts of the human body, from complete body systems to discrete organs, skin to bone, the visible to the microscopic. As the public face of the household anatomy studio, Anna Morandi offered a very different spectacle from the melodramatic scene performed annually in the Archiginnasio Anatomy Theater and analogized in the Anatomy Museum.

Morandi's expert lessons transported the anatomized body from the ghostly gloom of the Carnival Dissection into the clear light of the scientific laboratory. Hers was, of course, more than a conventional anatomy laboratory. She and her husband practiced anatomical science and physiology by means of their scrupulous dissections of hundreds of cadavers, written analyses of their findings, and engagement with dominant anatomical theories. In addition, they produced and circulated new knowledge of the body for specialists and amateurs alike by means of the meticulous and incorruptible wax anatomies they sculpted. Their muted aesthetic strictly inclined to the commandments of empirical science, privileging verisimilitude over the human passions. It was an art of science at odds with locally authorized visions of the body and thus—unlike Lelli's heroic écorchés—consigned to the periphery of Bologna's cultural establishment. The remainder of this book describes the work that took place in Bologna's "other" anatomy theater, and its varied influence in Bologna and beyond.

2

Professing Anatomy

Professione: *Definiz*: §. I. Per Esercizio, e Mestiero. Lat. *ars*. Gr. Τέχνη.
Definiz: §. II. Per Solenne promessa d'osservanza, che fanno i regolari. Lat.
*professio. Gr. προσομολόγησις.

Professore: *Definiz*: Che professa. Lat. *professor*. Gr. Διδάσκαλος

Esempio: But. Par. 24. 1. Colui è professore nella scienzia, che è di quella
approbatore, e può in quella approbare, e affermare per la lunga pratica,
ch'egli v'ha.

Esempio: Liv. M. Come professori di quella santissima filosofia.

    Vocabolario della Crusca (1733)

Professors will take care not to undertake in their work any scientific
study or discourse that is appropriate to the kind of lessons [given by]
instructors at the Public Studio, being required to direct their scientific
practice principally toward the performance of observations, operations,
experiments and other things of similar nature.

    Le Costituzioni dell'Istituto delle Scienze (1711)

Professing Anatomy

With the birth of the Institute of Sciences in Bologna in 1714, the significance
of the title "professor" (*professore*)—and of the associated action "to profess"
(*professare*) expert knowledge within an established branch of learning—
underwent momentous change. The founder of the institute, General Luigi
Marsili, defined his ideal *professore* against the traditionalist lecturer (*lettore*) of
the Bolognese Studium. The authority of the new *professore* no longer issued
from his skill as a channel of received theories. Rather, he secured his title
on the basis of his competency at focused empirical demonstrations within
the context of the scientific laboratory. In Marsili's words, he was to "find

himself in his own room, and there work on subjects relevant to his expertise . . . lovingly giving demonstrations to all those who wished to learn, be they citizens or foreigners."[1] By means of the practical application of his specialized knowledge and the concrete nature of his scientific results, the *professore* of the institute directly served the *pubblico benefizio* (public well-being) and thus merited public funding for his work.

While markedly revised, the new usage of *professore* did not dispatch entirely the ancient denotation of one who overtly declares his religious belief or membership in a religious order. Nor did it withdraw from associations with *professione* (profession), either as applied knowledge in one of the three great divisions of learning, "Divinity, Law and Physick," or as vocation, art, or paid occupation. Instead, it redeployed these connotations within the context of the experimental laboratory of the Enlightenment *House of Sciences* at the heart of the modern nation-state.

By midcentury, the Bolognese *professore* was inculcated in the core principles and praxis of European Enlightenment scientific culture that Marsili had successfully framed and put into effect in his native city. The Institute of Sciences professor overtly and pragmatically affirmed his belief in the truths revealed by experimental science, truths for which he solemnly toiled in his laboratory to the advantage of both the public good and his purse. It was in exactly this mode that Anna Morandi and Giovanni Manzolini understood and performed their work. Following is a reconstruction, to the extent possible from Morandi's anatomical notes and the couple's wax anatomies, municipal documents, and a critical eyewitness account, of how they fulfilled the principles and ideals of experimental science as *professors* of anatomy to "citizens and foreigners," specialists and amateurs, within their household laboratory.

To the Marrow of Things

Together in their home, the couple dissected a vast number of cadavers, more than a thousand by Morandi's own estimation, obtained primarily from the indelicately named Ospedale di Santa Maria della Morte (Hospital of Saint Mary of Death) in the city center, where the indigent sought care and the city mortuary was housed. Their method of anatomical discovery was revealed in the writings of visitors to the studio, but even more clearly in Morandi's anatomical notebook. By following the syntax of her inventory and her analysis of the anatomical structures she expressed in writing and in wax, it is possible to approximate the couple's methodological approach.

In her series demonstrating the anatomy of the pharynx, Morandi begins in the First Table by replicating the intact structure as seen *in situ* from a posterior, lateral position, in order to show the surrounding muscle groups and their points of attachment.[2] Because the pharynx is a muscular tube in the neck that forms part of the digestive and respiratory systems and aids in vocalization, Morandi endeavored to convey the distinct and integrated structures that contributed to these different functions. In the second display table, she moves progressively as in dissection by opening up the pharynx, still *in situ*, in order to show what she designated its three main cavities, superior, middle, and inferior (the nasal cavities, mouth, and esophagus), and related structures, all of which support respiration, digestion, and voice. Table Three, however, shifts from a contextualized to a particularized, anatomical view of the pharynx attached to the tongue, extracted from its "natural site" and exhibited in two separate figures. Her aim in these specific models was to detail the various strata, external to deep, of fascia, muscle, and mucous membrane, of which the pharynx is composed. The Fourth Table returns to a contextualized, summative study but from the anterior view, in order to show the relationship of the pharynx to the larynx, tongue, jaw, and other integrated parts. The comprehensive and restricted views of the pharynx and related structures in the series indicate the performance of multiple dissections. Notably, in her companion notes to the fourth display table, Morandi also details her discovery and the proper dissection of two minute "staphyline muscles" of the palate (labeled XX on the wax figure and in her notes) that, she asserts, Giovanni Battista Morgagni (1682–1771), the acclaimed professor of anatomy at Padua, misidentified as the "azygos" (meaning one, not of a pair) muscle of the uvula. She writes:

> The Staphyline or epistaphylines . . . are two small muscles that appear to be a single one, but are divided in some subjects along an extremely fine white line. They are attached to the border of the palate and run along nearly the entire length of the uvula. They are still called the Azygos of Morgagni for having been identified by him as a single one.[3]

Morandi's paradigmatic anatomy of the pharynx replicated the work of real scalpels and probes as well as the method of discovery by means of multiple dissections, from shallow to deep, of the organ's substantial and minute components, and the surrounding strata of fascia and related structures. As will be discussed at greater length in chapter 6, Morandi and her husband clearly practiced a method of dissection that, in contrast to today's methods of anatomical study and dissection, was defined by a systemic as opposed to a regional approach. It involved the isolation, study, and removal from the cadaver of discrete anatomical systems, such as the urogenital system, for separate analysis and dissection. The system would be extracted as a unit and examined in its entirety on the dissection table in order to see clearly the form, size, and shape of each organ or structure, and its origin, course, distribution, and function in relation to the other structures within the system. Discrete organs or structures of the system would then be bisected and studied, exactly as they were represented in wax on Morandi's wooden display tables.

It must be remembered, however, that Morandi's representations of the dissected body were also extra-experiential. What she conjured in wax was not what she saw and touched on her dissection table, but was instead an ideal vision of sentient flesh and animate bones. Her wax figures mediated very literally between death and life. They were cast from the dead and evoked the living. Her allied notes and waxworks rendered the animate pharynx in full and from varying perspectives within and outside the broader anatomical context in order for students and amateurs to understand thoroughly its standard placement, form, and physiology. On the authority of her more exacting scalpel, she also engaged and even boldly contested prevailing anatomical theories by such masters as the aforementioned Morgagni.

It is noteworthy that the method of wax modeling she and her husband employed to reiterate views of the body they encountered in the course of dissection was, in fact, the near-perfect reverse of their method of anatomical discovery. In modeling, they systematically reconstructed or, more appropriately, artfully fabricated, often directly on the actual skeletal core *stratum super stratum* of living muscle, organs, nerves, arteries, veins, and other intricate structures, upward to the external vestment of skin that they had cut through and dismantled layer by layer and piece by piece in the bodies of the dead. Theirs was thus a rich methodological double vision of the body unbound and remade.

Morandi's Tools

Another important source of information about the practical features of the Morandi-Manzolini anatomical practice is found in an appendix to Morandi's anatomical notebook. On these last pages of the manuscript appears a catalogue of Morandi's own dissecting and scientific instruments and sculpting tools, which were acquired by her patron Count Ranuzzi and sold after her death to the Institute of Sciences.[4] Listed in the catalogue, among other tools, are

a large iron saw
a small saw with a white handle
a curved knife for scraping bones clean
a fistula with a silver mouthpiece for introducing air

a small knife curved like a sickle
a pair of curved iron scissors from England
small files with handles
a small brass hand drill with five prongs
a marble millstone with a pestle of antique green
a steel rule divided by 270[5]
a cowl to protect against the stench of the cadaver
scalpels
forceps
iron probes
a table vice
a spatula for plaster
whetting stones
a small microscope

Anatomy of the Forearm

From Morandi's notes and collection of wax figures, the testimony of visitors to the studio, and this list of her dissecting and modeling tools, it is thus possible to conjure an image of Anna Morandi and Giovanni Manzolini in their home studio, cowled against the seepage and stench of the human viscera arrayed on their wooden dissection table (marble would no doubt have proved too costly). We picture them cleaning and depilating the muscular male forearm just retrieved from the nearby hospital *of the dead*. This "fresh part" would have served as the archetype to inaugurate an extensive ten-figure series on the limb.

With no means to preserve cadavers, dissection of the forearm necessarily occurred during the colder seasons and even at the coldest nighttime hours. In the dim glow of torchlight and with a steady fire to warm their fingers and take the sharpest chill off long hours of concentrated cutting and probing, they would prepare this and the many other arms to be used in the sequence.[6] The progressive dissection and representation of the layers and related structures of each limb or organ required the use of multiple body parts; other newly severed arms would thus be at the ready for dissection and casting.

Morandi and her husband demonstrated the fleshy, venous, and tendinous structures of the forearm by sculpting wax directly on actual bones. Indeed, the first figure was the only one of the series cast entirely in wax. In the ten-part display of the forearm, the couple therefore dissected at least nine arms and presumably many more to produce their various, ideal views of the anatomized appendage (figs. 27–28).

From the intact, depilated forearm, a plaster model would be made from which could be discerned through the skin the superficial muscles, tendons, bones, and veins, including, as Morandi writes in her notebook, "the external

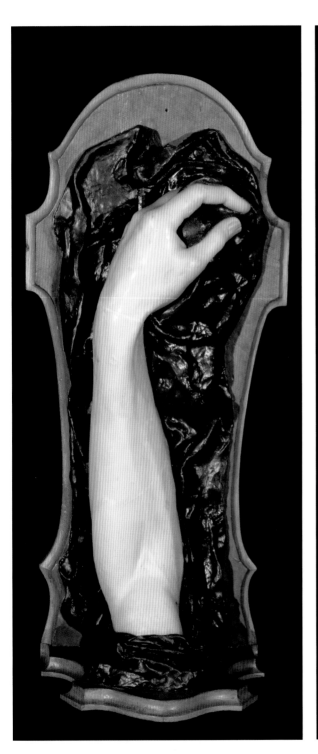

FIGURE 27–28. Anna Morandi and Giovanni Manzolini, forearm, wax and bone. Courtesy of Museo di Palazzo Poggi, Università di Bologna.

radial vein, the external cubital vein, the cephalic vein, the basilic vein, and the median vein, the cephalic vein of the pollux, and the salvatella vein." To produce the cast, Anna and Giovanni would first have smeared the prepared arm with animal fat so that it could be separated easily from the plaster. They would then have encased the severed forearm in thick layers of molding material. Skilled casting was among the most important stages of the design process, because the cast frequently served as a negative mold for subsequent reproductions. For example, the couple made numerous copies of their series on the sense organs for patrons throughout Europe, including the Procurator of Saint Mark's in Venice, Alvise IV Mocenigo, or, more probably, his wife, Pisana Corner, who was a devotee of anatomy, and the Royal Society of London.[7] It must be noted, however, that although ready-made plaster casts facilitated the production process and lowered costs, they would often be deformed or destroyed by the harsh casting process itself. Thus, the entire procedure, from acquisition and dissection of the cadaver to casting in plaster in wax, would be repeated.

Morandi's notes indicate that the second figure in the series showed the forearm with its superficial "coverings removed, and here one notes the muscles and parts that remain visible." To dissect what lies just under the skin, the anatomist-modeler must assertively grasp forceps and scalpel in each hand to reflect the skin of the arm with a lengthwise incision that reaches to the wrist, by which the skin can be peeled back with forceps or even fingers and separated from the subcutaneous flesh with careful flicks of the surgeon's knife, just as a skilled butcher detaches the puckered skin of a plucked chicken. Through slow, painstakingly applied force, the faint and fragile subcutaneous nerves and veins may be left intact for replication in wax. The dissecting partner might have assisted by readying extra knives and scalpels on the whetstone, or by undertaking the simultaneous dissection, at once intricate and strenuous, of the tendinous hand. The modeling of other parts and organs required a more refined and elaborate process of preparation as well as the use of precision instruments. The series on the eye—from the whole, animate eye within its socket, to the extrinsic muscles and, indeed, invisible ducts—was, for example, the result of many dissections, and called for the use of a microscope and wax injections.

Once the plaster had set and the limb could be removed, the modelers would then coat the interior of the cast with soap to seal it against the hot wax and to aid in the removal of the wax figure once it had hardened. Although we do not know their secret recipes for making wax, Morandi and Manzolini would have used the same basic ingredients combined since the first documented sculptors' *bozzetti* (notes) of the early sixteenth century. As the restorer Phoebe Weil records, the oldest known recipe for wax in modeling and bronze casting appeared in Pomponius Gauricus's *De Sculptura* of 1504.[8] His compound consisted of "three parts of beeswax mixed with enough oil or tallow to make it soft and pliable, to which was added one part of pitch [*pece*] softened with turpentine."[9] In "Sculpture," the ninth chapter of his *Lives of Artists*, Giorgio Vasari adds further details about the ingredients and their benefits for making modeling wax:

> In order to make it softer, you should add a bit of tallow, turpentine and black pitch, of which the tallow serves best to make [the wax] pliable, and the turpentine terse in itself, and pitch gives it a black color as well as a certain firmness such that it becomes hard after it is worked to completion.[10]

Notwithstanding the invariability of the basic amalgam from the sixteenth century to the eighteenth, even slight changes to the recipe could radically alter the strength and resilience of sculptures over time. Several eighteenth-century authors note, in fact, that Giovanni Manzolini improved on Ercole Lelli's recipe for making wax, and Morandi far surpassed both Lelli and her husband with respect to the naturalism and vibrancy of her colors as well as to the resistance of her wax compounds to cracking and decay.[11]

Working in tandem, Morandi and her husband would thus have undertaken the delicate process of filling the cast by successive layers with their special formula of hot, tinted wax. In the case of a smaller, intact organ, such as the tongue, the eye, the heart, and the kidney, they would likely have poured the wax through a hole at one end of the cast, which would then have been plugged until the wax hardened. Ingenious procedures were invented to facilitate the release of the wax figure from its cast. Most often, the cast was strategically cut into several pieces, which were fixed together by binding or by wooden or cork pegs.[12]

Once the waxen arm had been removed from the cast, the finishing stage of modeling would begin. Using slender scalpels and fine brushes dipped in turpentine, any surfeit of wax would be removed. Morandi, who would win wide acclaim for her precision in both dissection and wax modeling, manipulated hot iron tools to accentuate intricate details, like the graceful, horseshoe shape of the thumbnail (figure 29). She would then smooth and polish the arm to a glossy shine. In the case of the more elaborate figures of the dissected arm, she would use heated scalpels to carve anatomical minutiae and to add such components as naturalistically colored wax vessels and veins. She frequently used colored silk to indicate veins, and paint to accentuate fine details as well as to mark each component with identifying symbols (fig. 30). These symbols corresponded to the index and explanations in her companion anatomical

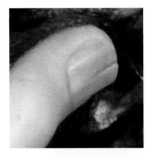

FIGURE 29 (LEFT). Anna Morandi, forearm (fig. 27), wax; detail. Courtesy of Museo di Palazzo Poggi, Università di Bologna.

FIGURE 30 (BELOW). Anna Morandi and Giovanni Manzolini, forearm (fig. 27), wax and bone; detail. Courtesy of Museo di Palazzo Poggi, Università di Bologna.

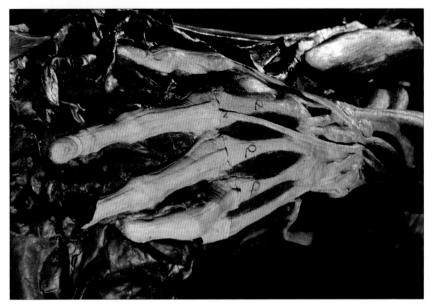

notebook that, apart from serving for Morandi's studio lessons in anatomy, could be used by her students and patrons to identify and study the anatomical figures independently (fig. 31).

The forearm, as with nearly all the figures in the collection, was fixed to a decoratively carved and painted wooden display table, whose form varied according to the size and shape of the body part demonstrated. Many of the *tavole* were octagonal, but most of the figures in the series on the arm were attached to an L-shaped, two-piece table, with the severed bicep resting on the short base and the extended forearm and hand stretching the length of the upright rectangular board. Morandi and her husband placed the entire forearm within a long crimped cloth soaked in a dark-colored wax that was attached to the table. The waxen wrap acted in this and other displays as a graceful curtain backdrop to the display of the anatomized part; but here it also served as a prop to emphasize the activated skeleton, muscles, and tendons of the liv-

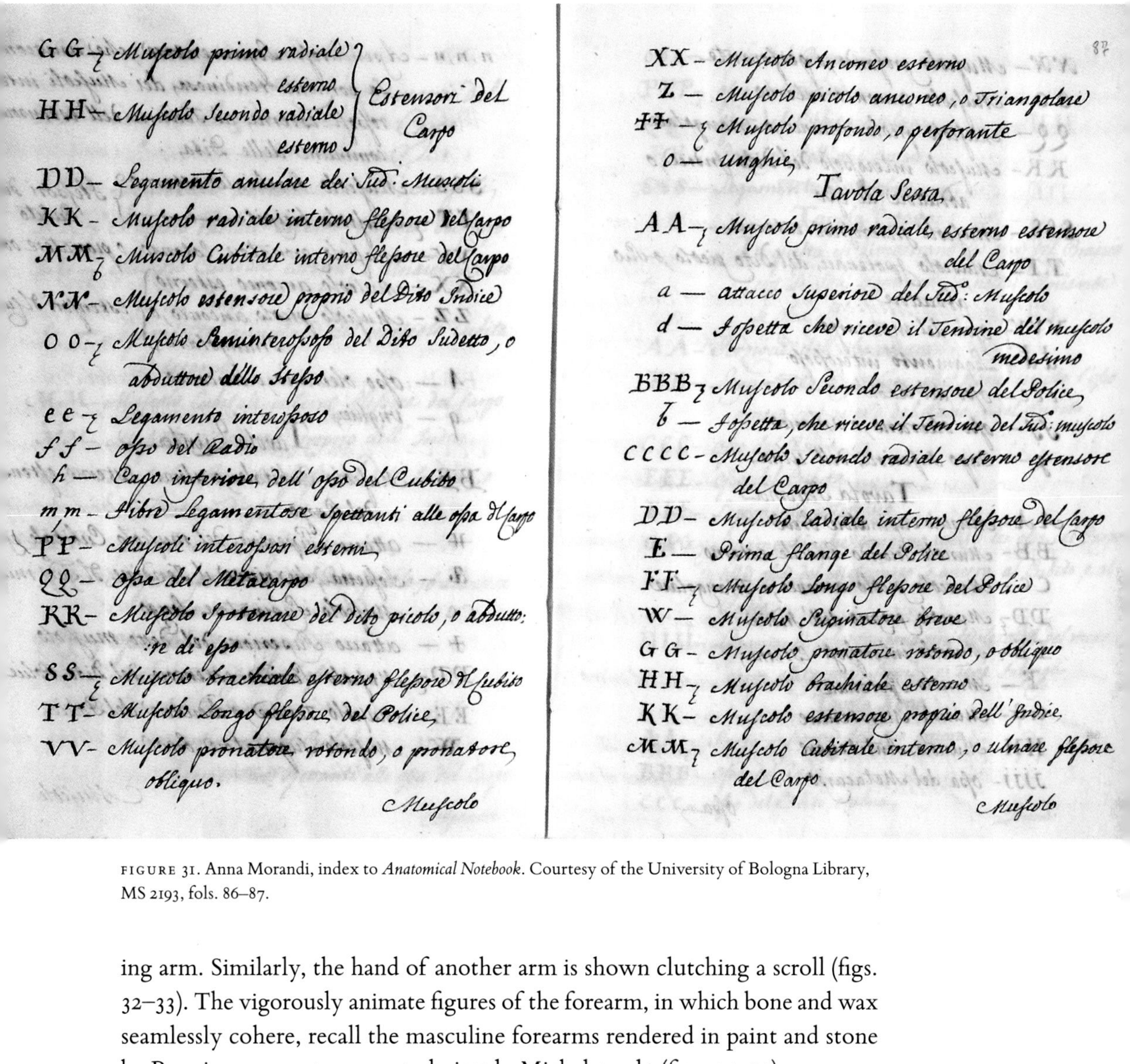

G G - Muscolo primo radiale
    esterno   } Estensori del
H H - Muscolo secondo radiale   Carpo
    esterno

D D - Legamento anulare dei sud. Muscoli
K K - Muscolo radiale interno flesore del Carpo
M M - Muscolo Cubitale interno flesore del Carpo
N N - Muscolo estensore proprio del Dito Indice
O O - Muscolo seminterossoso del Dito Sudetto, o
   abduttore dello stesso
e e - Legamento interossoso
f f - oso del Radio
h - Capo inferiore dell'oso del Cubito
m m - Fibre Legamentose spettanti alle osa d'Carpo
P P - Muscoli interossosi esterni
Q Q - osa del Metacarpo
R R - Muscolo Ipotenare del dito picolo, o abdutto:
   re di eso
S S - Muscolo brachiale esterno flesore d'Cubito
T T - Muscolo Longo flesore del Police
V V - Muscolo pronatore rotondo, o pronatore
   obliquo.
      Muscolo

X X - Muscolo Anconeo esterno
Z - Muscolo picolo anconeo, o Triangolare
Ŧ Ŧ - Muscolo profondo, o perforante
o - unghie

    Tavola Sesta
A A - Muscolo primo radiale, esterno estensore
    del Carpo
a - attacco superiore del sud. Muscolo
d - fosetta che riceve il Tendine del muscolo
     medesimo
B B B - Muscolo secondo estensore del Police
b - fosetta, che riceve il Tendine del sud. muscolo
C C C C - Muscolo secondo radiale esterno estensore
    del Carpo
D D - Muscolo radiale interno flesore del Carpo
E - Prima flange del Police
F F - Muscolo Longo flesore del Police
W - Muscolo Supinatore breve
G G - Muscolo pronatore rotondo, o obliquo
H H - Muscolo brachiale esterno
K K - Muscolo estensore proprio dell'Indice
M M - Muscolo Cubitale interno, o ulnare flesore
    del Carpo
      Muscolo

FIGURE 31. Anna Morandi, index to *Anatomical Notebook*. Courtesy of the University of Bologna Library, MS 2193, fols. 86–87.

ing arm. Similarly, the hand of another arm is shown clutching a scroll (figs. 32–33). The vigorously animate figures of the forearm, in which bone and wax seamlessly cohere, recall the masculine forearms rendered in paint and stone by Renaissance masters, most obviously Michelangelo (figs. 34–35).

Morandi and Manzolini guided and supplemented their hands-on anatomical discovery of the body with Vesalius, Cowper, Valverde, Valsalva, and other master anatomical atlases in their home archive. Their waxworks were based therefore on the critical, combined authority of body and book. The relevant pages of one or more atlases likely rested open at arm's reach for consultation. Large ceramic collecting pots would have lined the studio floor beneath the table for the skin, the layers of oily, yellow adipose, and other sected oddments and refuse.

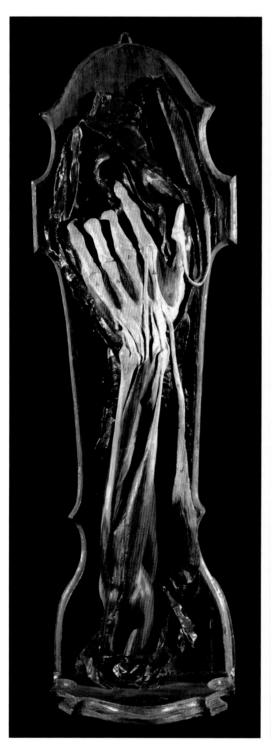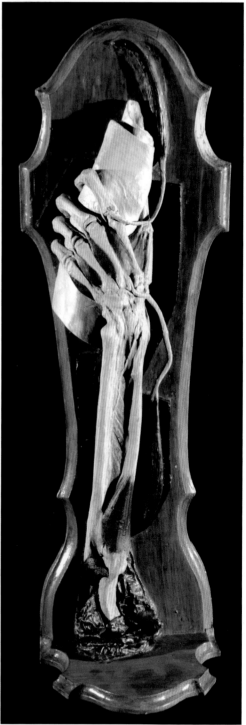

FIGURE 32–33. Anna Morandi and Giovanni Manzolini, muscles and bones of the forearm. Courtesy of Museo di Palazzo Poggi, Università di Bologna.

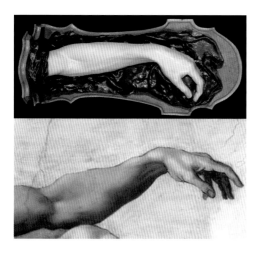

FIGURE 34–35. Anna Morandi, forearm, wax and bone. Courtesy of Museo di Palazzo Poggi, Università di Bologna. Michelangelo, forearm of Adam; detail from *The Creation of Man*, Sistine Chapel, Rome.

The Household Studio

Morandi and her husband only demonstrated human anatomy within their home, but this was not unusual. The household school and laboratory was by this time a well-established domain of Bolognese academic life. Private schools in the homes of professors had begun to flourish in Bologna from the second half of the seventeenth century, at the exact moment when religious and municipal authorities imposed an exclusively theoretical method and curriculum at the university and prohibited instructors from departing from authorized texts and interpretations.[13] More than any other subject matter, in fact, human and comparative anatomy and anatomical dissection took refuge in household studios from the repressive hold of faded methods and texts. Stymied by the suppression of new authors and empirical analysis, as well as by the rarity of public dissections, the study of gross anatomy and dissection survived and indeed made crucial progress within the familial walls of Bolognese scholars. Although perhaps shocking for us to envisage, in their home laboratories university professors gave paying students hands-on instruction in human dissection, animal dissection, and animal vivisection.

Marcello Malpighi's experience as a young medical student epitomizes the home school tradition and its importance for the development of a new class of modern anatomists. As one of nine members of the elite Coro Anatomico (Anatomical Chorus) founded in 1650 by the professor of medicine Bartolomeo Massari, Malpighi received in his professor's home his first exposure to the most important medical texts of the time as well as to the latest medical methods.[14] By his own telling, his passion for anatomical science grew out of his experiences in the household anatomy theater, where his teacher privately conducted anatomical dissections of his own election. Moreover, frequent dissections were conducted by Signor Giovanni Battista Capponi, with the assistance of Signor Golfieri, Fracassati and other members of the Coro Anatomico on various live animals and, with the availability of executed criminals, on human beings, as a result of which I became most intrigued with this profession.[15]

Malpighi alludes to the distinct advantages of the household anatomy over the annual Public Anatomy. As a rule, the private anatomy lesson permitted far more intimate views of the dissection, hands-on knowledge of the parts and organs of the body, and frequently the use of specialized instruments, including microscopes, bellows for blowing air into the lungs, and wax-injecting tools.

In contrast to Malpighi's day, however, at the very moment when Morandi and Manzolini began their household anatomy practice, the study of practical anatomy was undergoing a crucial shift upward in status, approaching the privileged place held by such scientific disciplines as chemistry and physics, which were practiced within the Institute of Sciences. Pope Benedict XIV's award of a new "Chair for Life" of surgery to Pier Paolo Molinelli is a clear sign of the increasing prestige of anatomical study and dissection.[16] However, Molinelli's letters in 1743 to Flaminio Scarselli, secretary to Bologna's ambassador to Rome, make plain the surgeon's desire to teach surgery in a more fitting and prominent place than the Hospital of Santa Maria della Morte—that is, within a designated room of the institute, and specifically in the room displaying Anton Maria Valsalva's celebrated collection of tools and anatomical specimens. Molinelli's letter efficiently summarizes important conventions governing the university teaching of practical surgery and anatomy from cadavers at this time, as well as the deficiencies of those same conventions for ambitious young professors:

> One might argue that practical surgery should be taught as usual in the hospitals, in accordance with the Pope's order. But one might also answer, why not in the Institute? Practical anatomy is taught in the public schools and in the Home of a Professor throughout the year. Because of this should we neglect to establish a magnificent school within the Institute? There are certain professions that are never taught or cultivated enough. You might then counter that the Institute is not a place for cadavers. But this is not an issue. . . . If one were to demonstrate operations on the eyes in the Institute for example, that would require that a head alone be brought into the surgical theater and not the entire cadaver. And I don't doubt that with the establishment of a school of anatomy in the Institute, it would not be necessary to bring there entire cadavers but instead some parts of the same.[17]

Molinelli indicates that Bologna's three prime venues for anatomical and surgical demonstrations with the use of cadavers were the "public schools," in other words, the university, where the annual Public Anatomy was undertaken in the Archiginnasio Anatomy Theater; the anatomy and surgical laboratories in the hospitals of Santa Maria della Morte and Santa Maria della Vita; and the home studio of a professor, who, Molinelli tells us, customarily offered university students a comprehensive, yearlong course in practical anatomy.

Yet while hands-on study of the body was gaining academic prestige, numerous documents of the Assunteria di Studio (the Senate body that supervised the administration of the university) also reflect widespread disquiet among political and university leaders over Bologna's dearth of adequately trained anatomists, especially those willing and capable of performing the celebrated annual Public Anatomy. On 16 June 1750, the Senate received the following report:

> Members of the Assunteria, recognizing as their duty and their office to place before the eyes of the Senate all that might contribute to the preservation of the luster of this our most celebrated University must, under no circumstance, refrain from representing to the Most Illustrious and Excellent Senators the state in which the current class of practicing anatomists has been reduced and the reasonable apprehension in which they find themselves, who without a solution to the scarcity of these same anatomists and to the selection of new lectors to make up for the current decline, in very few years could be too few to sustain the weight of an operation of such great prestige and fame for this University.[18]

In their formal response to the problem, university officials stipulated in writing the necessary qualifications of the practical anatomist, who "must be a good lecturer and capable in physics, mathematics, medicine and especially practical anatomy,"[19] which was most fittingly learned under the direction of a "Lector of Anatomy who teaches practical anatomy in his home."

By the mid-eighteenth century in Bologna, therefore, experienced anatomy instructors, especially outside the ranks of the aged emeriti, were in limited supply. Insufficient opportunities existed for comprehensive, practical anatomy training in the two hospitals designated for this purpose. As we have seen, the Carnival Dissection served the cause of civic glory and pride, among other more esoteric functions, rather than practicable student instruction. The age-old stigma of dissecting the dead continued to influence the teaching of practical anatomy and surgery, which was barred from the Institute of Sciences and consigned to subterranean hospital laboratories, the annual Public Anatomy, and, most commonly, professors' private homes. The household anatomy studio thus continued to be, much as Marcello Malpighi had described it a hundred years before, the principal site of educational dissections and anatomical training. It is within this distinctive cultural and institutional context that the work and the authority of the Morandi-Manzolini studio are rightly viewed.

Teaching the "New" Anatomy

Despite the scarcity of concrete data about who trained with them, Morandi and Manzolini unquestionably taught an abundance of medical students as well as avid recreational anatomists. Because of their full access to cadavers and body parts and the vast number of dissections they performed, the couple could provide extensive practical training in human anatomy to students, who clearly would have been hard pressed to find similar instruction elsewhere. Moreover, by the early 1750s, Morandi and Manzolini had already achieved local and international fame as anatomists and anatomical modelers and enjoyed frequent visits by Grand Tourists, many of whom commented on the pedagogical focus of their practice and the numerous students in their charge. It is reasonable to assume therefore that their studio would have enjoyed high standing among budding and established medical professionals, who would not only have received specialized instruction from two prominent experts in the field, but could potentially enhance their own status in the profession by association with the couple and their celebrated studio.

What most set the Morandi-Manzolini anatomical practice apart was, of course, their use of their expansive and unparalleled collection of wax models for teaching anatomy. According to their contemporary biographer, Luigi Crespi, Giovanni Manzolini specifically created his models for the benefit of "native and foreign youth" desirous of training in anatomical science.[20] Medical students who frequented the studio enjoyed the unique advantage of studying the myriad parts and structures of the body during any season and at any time of day without the inconvenience of the cadaver's putrefaction and without fear of contagion from these same moldering parts. In their illustration of an extra-experiential reality of the anatomized body devoid of the disorder and decay inherent in human dissection, the couple's waxworks literally embodied new anatomical knowledge. They cleared away the messy obstacles to sight and understanding encountered in dismembering the dead and at the same time better elucidated the complexity of the human form. They were able to emphasize and isolate discrete structures, including very fragile and minute body parts that would have been lost or corrupted in an actual anatomical specimen. Their models more effectively reconciled theory and phenomenon by offering clearer and more comprehensive views of the anatomized body, more exacting analysis, and an exhibition of the body's parts in some ways superior to an actual anatomical preparation. The models thus represented and obliged new standards for studying the anatomical body.

Speculum

Equal in novelty and intrigue to the wax models was, of course, the woman anatomist who demonstrated them. The firsthand account by the cosmopolitan physician Giovanni Bianchi of his visit to Morandi's home studio provides a glimpse of one of her actual anatomical demonstrations, but one in which authentic traces of her presence and scientific method may be discerned. On the occasion of Bianchi's visit, Morandi began with a comparative analysis of her vast collection of skeletons, ranging from fetal to adult development. Although Bianchi gives no description of the studio itself, it would necessarily have been large enough to house and display this series as well as the couple's complete collection of hundreds of figures to which Morandi would refer throughout her lectures. The wooden tables of myriad shapes and sizes exhibiting one or more wax figures may have hung on the walls or lined the shelves throughout the room, in the well-known tradition of the curiosity cabinet. Of course, the studio would also have had to accommodate a limited viewing public.

Morandi's study of the skeleton was the most extensive in her oeuvre, elaborated in countless pieces that covered 22 tables, and were described in 62 pages of her notes. One display table alone, for example, exhibited 30 figures. Unless she devoted several hours solely to her lecture on osteology, ranging from the fetal stage beginning at one month, to the immature skeletons of small children in progressive order to the fully developed skeletons of male and female adults, she would have had to limit her presentation to certain figures and themes. She undoubtedly stressed, however, the development and transformation of the skeletal structure from the smallest human specimen in evidence to the most mature. It should be noted that Morandi's studies of the skeletal structure comprised actual human bones uncovered during her dissections. No wax figures were involved here.

Moving from the inside out during Bianchi's visit, she then focused on the myology, or muscular structure, specifically of the arms and feet. (Earlier in this chapter, I described the series she created with her husband on the forearm, their method of dissection and representation, and her contestation in her notebook of established anatomical theories to which she would have undoubtedly referred in this portion of her lecture.) Bianchi also indicates that he heard Morandi lecture on her famed sense organs: the eye, ear, nose, and "organs of voice," although he makes no mention of her celebrated series

art than craft, more "head-work than handiwork," Lelli has been allotted a place, albeit relatively modest, in the annals of art history.[29] By contrast, there is no trace of Giovanni Manzolini. Beyond Freedberg's study of her, if Anna Morandi is mentioned at all she typically represents a tangential figure through her honorary affiliation with the Clementina Academy of Art. The distinction between art (Lelli's works) and non-art (Manzolini and Morandi's works) hinges not only, as Freedberg has discussed, on the "explicit opposition between exactitude and beauty," but also on the more esoteric dichotomy between the practicable and the sublime.[30]

Excluding such rare exceptions as Freedberg's study, the aesthetic and the science Morandi practiced in imaging the anatomized body has been largely ignored. Not only have critics relegated Morandi to the margins of Bolognese art history, as will be discussed subsequently, from her own day to the present time, she has been effectively consigned to the remote fringe of the canon of eighteenth-century scientists. Four recent studies of note, which focus extensively on the intersection of art and science in the eighteenth century as expressed in wax anatomical sculpture, spill pools of ink on the pioneering and spectacular work of the baroque artist Gaetano Zumbo and his "noteworthy" successors in anatomical display in Florence late in the century, Felice Fontana and his *ceraiolo* (wax modeler), Clemente Susini. Ercole Lelli's "founding of the Bolognese school" and critical innovations in ceroplasty are discussed at length in two of the studies. Morandi receives no mention in two of these and passing mention and misidentification as a student of Lelli's in the other two.[31] The high esteem paid by contemporary critics to the clamorous and eroticized visual pageants of human flesh and mortality represented in the baroque waxworks of Gaetano Zumbo, the epic écorchés of Ercole Lelli, and Felice Fontana's anatomical Venuses in contrast to the feeble acknowledgment of Morandi and Manzolini's dispassionate wax narratives, studied apposites to the living interior of the body, reveals the critical bias against a muted body aesthetics rooted in anatomical verisimilitude. Morandi did not retrofit her images of human anatomy into such traditional moralizing contexts as *The Fall, The Dying Gaul, The Plague,* or *The Triumph of Time.*[32] As her notes to the anatomy of the hand reveal, however, the drama she sought to represent was arguably more, not less, poignant than that of esteemed, canonical wax sculptors:

> We have shown two hands expressing those tender and bitter sensations that result from the embrace or touch of the palm . . . in natural accord with the qualities of the objects they touched. Here we therefore demonstrate the internal parts of the hand [that facilitate these reactions] by stripping away its coverings in order to see the muscles, tendons, nerves, fascia, etc.[33]

Morandi was well aware, of course, that the mix of the organic and the synthetic, the real and the artificial—human bones and beeswax, resin, oil paint, glass eyes, body and head hair—separated the wax effigy from traditional portrait media: paint and canvas, stone, wood, and bronze. The material and psychological distance established between the actual human subject and conventional portraiture disturbingly diminished with the biomorphic effigy in colored wax. As Ernst H. Gombrich has observed, the wax image "often causes us uneasiness because it oversteps the boundary of symbolism."[34] In contrast to its auxiliary function in bronze casting or marble statuary, wax here is the prime creative medium. An uncannily lifelike substance, malleable and compliant to such true-to-life features as hair and glass eyes, it can be fit to, be molded on, and intimately commune with the actual body being replicated. As George Didi-Huberman poetically explains, "this vegetal material that bees have 'digested' in their bodies and in a sense rendered organic, this material nestled against my flesh becomes *like flesh*. This is its subtlety, but also its sovereign *power*: everything in it—plasticity, instability, fragility, sensitivity to heat, and so on—suggests the feeling or fantasy of flesh."[35] It is the potent efflux of ambiguity, the ontological liminality of the wax model, which stands between flesh and fiction, life and art, the material and the divine that undergirds its psychological sway and can provoke deeply ambivalent responses in viewers.[36]

Yet Morandi, I would argue, aimed to transform into scientific satisfaction and visual delight the tense ambivalence awakened in viewers by the wax ex voto and naturalistic wax portraiture,[37] which were always redolent with death even as they imaged the living. Morandi's anatomical figures were a palpable means to knowledge of what is alive and at work beneath the skin. Revelatory of the interior of the person, her anatomical sculptures were, however, depersonalized archetypes, disconnected from biographical narrative or historical and moral context. She in fact created no écorchés or complete anatomical figures, always emphasizing instead partial views—arms in motion, a flowing urogenital system, darting eyeballs, even fractional vibrant faces—which were, however, deliberately and obviously synecdochical in their evocation of the whole anatomical and physiological context (fig. 36).

By means of her particular "making and matching," as Gombrich has described the artistic rendering of nature, Anna Morandi aimed above all to familiarize the unfamiliar, visualize the invisible not only of the anatomical body but of the relation of its form to its functions, its internal structures to its external physical experience. The predominant force of her images thus drew from their ability to instantiate new and useful knowledge of the body.

FIGURE 36. Anna Morandi, face, wax. Courtesy of Museo di Palazzo Poggi, Università di Bologna.

3

Re-casting

Oh degna che di Lei mai nessuna età possa tacere!

    LUIGI GALVANI, "De Manzoliniana"

True sciences are those which have penetrated through the senses as a result of experience.

    LEONARDO DA VINCI, *Trattato della pittura*[1]

The untutored female improviser (*improvvisatrice*), typically of humble birth, whose intrinsic brilliance in the arts and sciences revealed itself on impulse, was a common cultural trope of the Italian eighteenth century. A paragon of a distinctively feminine creative and intellectual virtuosity, the case of the *improvvisatrice* frequently impinged on the expansive debate across the peninsula about women's education, both for and against.[2] Like the priestess of the "poetics of enthusiasm," Corilla Olimpica (Maria Maddalena Morelli Fernandez, 1727–1800), who attained the summit of literary acclaim in 1776 as the only woman ever crowned poet laureate in Italy, Anna Morandi was depicted as having surmounted her modest social and intellectual station by virtue of serendipity and raw genius. Eighteenth-century biographers, in fact, repeatedly cast Morandi as an *improvvisatrice* of anatomy much as they deemed the poetess Corilla a kind of extemporaneous scientist. As told in the *Acts of the Arcadia Academy*, of which Corilla was the most famous (to some, infamous) member of her age, the *poetessa*, "ignited by wonderful inspiration and frequently changing her meter and her harmony and in an inexhaustible poetic vein, expounded on all the various scientific subjects she had been asked to address,"[3] including physics, metaphysics, and moral philosophy. Morandi's uncultivated genius in anatomy was likewise seen to have bloomed instinctively.

The highly influential five-and-a-half-page biographical profile of Morandi, published in 1769 by the artist and historian Luigi Crespi, represented Morandi's entry into the field of anatomical science as an accidental consequence of her devotion to hearth, God, and husband:

> With her vigilant attention to domestic affairs, her exemplary piety, and her constant encouragement of her studious, melancholy, pusillanimous, and afflicted husband, she was not only a source of every comfort and consolation, but with her talent and robust spirit she was also a source for him of aid and support.[4]

It was in this spirit of feminine compassion and self-sacrifice, according to Crespi, that

> she thought one day to attempt to work in the same profession of anatomy as her husband in order to assist and encourage him. And although she was nauseated and disgusted at first, after trusting in divine aid and forcing herself to overcome her repulsion, she began to dissect cadavers and to make discoveries, observations and incisions, finally becoming quite adept in this art.[5]

Her husband's professional and moral distress coupled with her own devoutness were thus seen to have aroused in Morandi an intuitive intellectual daring and a natural capability that carried her from a state of dedicated domesticity to one of expert competence in the rigors of anatomical dissection. With some coaching from her husband and by "domesticating herself" (*addimesticandosi*) to the fouler aspects of the study of anatomy, she came to cut apart cadavers, to sculpt anatomical figures, and to teach medical students in her home.[6]

Yet, in contrast to Corilla Olimpica's spectacular poetic art, Morandi's waxworks did not serve to minister her own glory, according to Crespi and his many followers, but her mate's. Morandi's self-sacrifice in the aid of her husband thus occasioned the emergence of her anatomical and artistic faculties. She was, therefore, a more perfect exemplar of the *improvvisatrice* than the *poetessa* Corilla herself, who was widely censured for abandoning husband and child in order to fulfill her artistic ambition. Echoing Crespi, the nineteenth-century biographer Michele Medici encapsulates Anna Morandi's altruistic rise:

> [Giovanni Manzolini] found timely and ready support in his woman, who, tenderly loving him as she did, and fearing that he might abandon his esteemed works, or continue with less desire and alacrity of spirit, devoted herself entirely

to his comfort, not, however, with sweet and gentle words, but by herself becoming an anatomical sculptor. She enhanced her study of design . . . to our as-
tonishment, by drawing herself close to the cadavers and with a virile and strong
spirit and incredible constancy dissecting them and uncovering and illustrating
their most obscure parts.[7]

Medici's praise of Morandi also highlights the critical and exceptional role of
the body of the woman anatomist and modeler in executing her art, a role
that subtly matches key attributes of the improviser's poetic performance.
Numerous firsthand accounts recall that, once in possession of her impromptu
theme, the *improvvisatrice* would begin to dance, to sing, and to extemporize
rhythmic, poetic lines, eventually entering a state of heightened concentration
in her advance toward pure poetic expression, unbound and unmediated by
university erudition, rhetorical schemes, paper and pen. In his popular trav-
elogue, John Moore describes a characteristic poetic performance by Corilla
Olimpica:

> After much entreaty, a subject being given, she began, accompanied by two
> violins, and sung her unpremeditated strains with great variety of thought and
> elegance of language. The whole of her performance lasted above an hour,
> with three or four pauses, of about five minutes each, which seemed necessary,
> more that she might recover her strength and voice, than for recollection; for
> . . . nothing could have more the air of inspiration, or what we are told of the
> Pythian Prophetess. At her first setting out, her manner was sedate, or rather
> cold; but gradually becoming animated, her voice rose, her eyes sparkled, and
> the rapidity and beauty of her expressions and ideas seemed supernatural.[8]

Eyewitnesses frequently commented on the extraordinary stamina required
for these performances, which at times ended in physical collapse. Observers
also emphasized the powerful affect of the performance on their own bodies
and states of mind. The body of the improviser, which was inseparable from
her poetic art, was thus a spectacular instrument and an instrumental spectacle. Indeed, on the rare occasion when the female improviser's verses were
published, they commonly met a subdued reception, because the elemental
creative part played by her body was inevitably lost in translation.[9]

Although Morandi never danced or swooned from anatomical enthusiasm
at the dissection table or in her creation of new bodies in wax, her body was
seen, like that of the female improviser, as the font, the vital medium, and
the translator of her innate creative intelligence. The tributes of Francesco
Maria Zanotti, Luigi Crespi, and Jacopo Bartolomeo Beccari, among others,
laid emphasis on the intuitive talent and the "virile and strong spirit" that
enabled her to undertake both the gruesome work of human dissection as
well as the refined art of refabricating bodies in wax. In June of 1754, Zanotti,

tural authority and favor from the pope and the pope's powerful surrogates among the Bolognese aristocracy. In Crespi's view, Lelli had procured this favor by shamelessly appropriating the work of other artists, most egregiously that of Manzolini, the more talented and proficient assistant on the pope's museum commission. Apparently, Crespi also bore a personal grudge against Lelli, who, as Prince of the Clementina Academy of Art (a post secured with the pope's intervention), had blocked Crespi's admission to the academy.[19] Yet, as we have seen in chapter 2, far more contemptible for Crespi and a broad swath of like-minded Bolognese artists, who adhered to an entrenched baroque aesthetic and to traditional distinctions between high and low art, was Lelli's propagation through his anatomical waxes of a demeaning "pragmatic conception of art."[20] Lelli's greatest sin, for which he was only "deserving of derision,"[21] was that of upturning the traditional hierarchy of the arts, wherein the mechanical rests on the lowest rung, and for thus diluting art's true esoteric object.[22]

Remarkably, this polemic continues to play out among today's historians, with some writing in vigorous defense of Lelli's modernizing impact on eighteenth-century anatomical design and others defending the validity of Crespi's accusations.[23] What historians have failed to account for, however, are the clear distinctions Crespi makes between the artist (Lelli and his disciples), who, he alleged, impersonated the scientist and counterfeited science in his art, and the honorable artist-cum-scientist (Manzolini), who bent artistic skill to the utilitarian tenets of a science in which he was genuinely proficient. In his biographical profile of Manzolini, Crespi actually highlights how the artist naturally veered from his original career path as painter and sculptor in order to pursue expertise in anatomy and anatomical modeling, in which he demonstrated that "supreme diligence, attention and practice in the subject" necessary to becoming one of its foremost "professors." [24] For Crespi, the most incontrovertible mark of distinction between Lelli's imposture, on the one hand, and Manzolini's honest capabilities on the other, was the approbation of illustrious scientists, including the Bolognese obstetrician and professor of surgery Giovan Antonio Galli and the surgeon and anatomist Pier Paolo Molinelli, for whom Manzolini and Morandi created wax anatomies for practical use in the instruction of medical students. Where Lelli failed both as artist and, in his capacity as Prince of the Clementina Academy of Art, as guardian of high artistic tradition, standards, and practice, Manzolini and Morandi soared at the practical handicraft (dissection and sculpture) of anatomical science. Simply put, Luigi Crespi appraised Ercole Lelli primarily according to aesthetic principles and Manzolini according to scientific ones.

Crespi, in fact, conferred unique standing in his *Lives of Bolognese Artists* on

Giovanni Manzolini and, by her association with Manzolini and extension of his work, on Anna Morandi as well. In his defense of Manzolini and his profile of Morandi, he tacitly conceived of a new class of artist that combined the experiential knowledge of the scientist and the artisan with the textual (literary and visual) knowledge of the scientist and the traditional high artist for the production of a new art of science. The material translation of knowledge that Manzolini and Morandi undertook set them apart within the broad class of artists.

Then and now, however, Morandi has been largely ignored or regarded as peripheral to the cultural quarrel over the purpose of anatomical wax design. In fact, she has never been mentioned as a protagonist in the dispute with Lelli.[25] Noted contemporaries of Morandi, Francesco Zanotti, Giovanni Fantuzzi, Crespi, and others, who extolled her anatomical talent, and those who took sides in the debate over the purpose of anatomical design, including Pisarri, Bianconi, and Algarotti, never mention Morandi as an actor in the controversy. Since then, critics and historians have also disregarded her part. That Lelli himself saw Morandi as a potent rival in the sphere of anatomical design is clear from the correspondence by, to, and about him during his lifetime. As we have seen, his letters to Marc Antonio Laurenti in Rome lay bare his anxiety over her public ascent at the very moment he was struggling to complete Pope Benedict's Anatomy Museum.[26] Undoubtedly, Morandi's perceived status outside the bounds of dominant social convention as an exceptional woman—an anatomical *improvvisatrice*—practicing her "natural" art within the confines of her domestic anatomy and modeling studio contributed to the virtually wholesale neglect of her role in redefining the method and purpose of anatomical design. This role began, of course, in partnership with her husband, a comprehensive and thus extraordinary partnership that laid the foundation for her solo dissections, castings from death, and demonstrations of the anatomy of the living body in wax and written word.

Lab Partners

It is difficult to ascertain the epistemological scope of Morandi's scientific marriage to Manzolini from their eighteenth-century biographers. As we have seen, contemporary Bolognese academicians frequently reiterated the same trite characterization of their partnership: Morandi's self-sacrificing devotion to her dejected husband was the critical, unforeseen catalyst for their side-by-side toil at the dissection table. A vision of the practical and theoretical aims of their work together has thus been blurred in a miasma of truisms and fractured from gaps and elisions in the primary documentation. Only the couple's written and wax studies of human anatomy provide a reliable, nuanced view of their partnership.

In 1746, Galli (fig. 37) commissioned Giovanni Manzolini and Anna Morandi to sculpt in wax the original twenty of more than 150 models of the placenta, the gravid uterus with its attachments, and other parts of the female reproductive system that Galli would use in his academic lectures and household school of obstetrics.[27] Manzolini and Morandi together prepared these figures, which conformed, as per Galli's directive, to Galli's own drawings based on his dissections of the female reproductive system, as well as to the noted obstetrical studies of François Mauriceau and Hendrick van Deventer.[28] Galli's was the first recorded commission received by the couple after they had opened their home studio, and one that would generate immediate interest in their work within and outside Bologna.

Giovan Antonio Galli's "museum," as he called it, of obstetrical wax figures contributed to the changing study and practice of obstetrics taking

FIGURE 37. Angelo Crescimbeni, Giovan Antonio Galli (1775). Courtesy of Museo di Palazzo Poggi, Università di Bologna.

place in Italy at the time. This was in no small part due to Morandi's and Manzolini's extensive theoretical and practical knowledge of human anatomy and their distinct mode of wax anatomical representation. At a time when Ercole Lelli was in the midst of completing the centerpiece écorchés of the pope's Anatomy Museum designed to enthrall Grand Tourists and to instruct Bolognese artists in a more accurate art of the nude, Manzolini and Morandi were specializing in the practical science of anatomy in order to create the most exact replicas yet seen of the anatomized body for use by anatomists and surgeons and for their own instruction of medical students. The couple contributed to the accretion of precision tools for the scientific reform and medicalization of obstetrics burgeoning at this moment in history.

With Morandi and Manzolini's wax obstetrical figures, as with such obstetrical instruments as the speculum and forceps that he had imported from France for his practice, Galli sought to elevate the scholarly and scientific authority of obstetrics, a traditionally feminine sphere of artisanal healing whose practices were suffused with mystery and rooted in oral as opposed to textual traditions and transmission of knowledge. Indeed, as Claudia Pancino has shown, "at the beginning of the eighteenth century the practice of assisting with childbirth was entirely entrusted to midwives, or, in their absence, to mutual aid among women."[29] The occupation of midwifery was, moreover, largely untouched by theoretical or empirical obstetrical studies.

However, as the century progressed, obstetrics, along with surgery, the two medical vocations traditionally held in lowest esteem, were increasingly professionalized and grew in stature within the sphere of the medical sciences. At Pope Benedict's behest, as we have seen, Pier Paolo Molinelli became the first chair of surgery in Bologna in 1742. Between 1732 and 1779, fourteen schools of obstetrics opened across the northern half of the Italian peninsula, most with the backing of local governments.[30] In Bologna from the 1740s, obstetrics and surgery began to pass from the practiced, muscular hands of midwives and barber-prosectors to the refined touch of physician-professors. Wax anatomy figures facilitated that passage. With facsimile uteruses, placentas, hearts, ears, eyes, and so forth, the physician-professor could teach literally without dirtying his hands.

Galli was a bridge figure at this transitional moment for both of these branches of practical medicine. For ten years he had worked rather ingloriously as a professor of surgery and a practicing obstetrician before being called to join Bologna's elite Academy of Science within the Institute of Sciences. Only after teaching obstetrics for twenty years to midwives and surgeons in his home studio did he win a place for his school within the Institute of Sciences, an honor conferred on him above all because of his astounding collection of anatomical models and modern instruments.[31] It is important to note, however, that Lelli frantically worked to keep Galli and his collection out of the institute because Galli's most prestigious obstetrical waxworks had been made by Giovanni Manzolini and Anna Morandi.[32] In this case Lelli was unsuccessful.

Although he taught obstetrics to increasing numbers of male surgeons within the institute, midwives remained a core constituent of Galli's in this prestigious site of science. Indeed, when the pope authorized purchase of the museum of some 170 wax and clay obstetrical figures and obstetrical instruments in 1757 and had these installed the following year in two rooms on the first floor of the institute, the street-level site was carefully chosen because it could accommodate a *portella*, a secret door by which the midwives could enter without being seen.

Notably, however, Giovan Antonio Galli did not work to subordinate or repress the profession of midwifery, as was the increasing trend across Europe during the eighteenth century.[33] It is clear from his devoted work with midwives that he instead aimed to educate them in safe and hygienic obstetrical methods in order to decrease the suffering and death rampant among birthing mothers and their newborns. Indeed, in his proposal to local civic leaders outlining the expansive reform of the practice of obstetrics in Bologna, Galli was unequivocal that surgeons trained to intervene in high-risk pregnancies should refrain from quarrelling with midwives and favoring some in the profession over others. He demanded that these surgeons instead show respect to midwives officially certified in obstetrics and openly support their methods of care.[34] Several years after Galli had opened his private obstetrical school, the French midwife Madame du Coudray would be pensioned by King Louis XV to provide a similar program of practical education and certification in safe birthing techniques to midwives across France.[35]

The sixty obstetrical lessons per year that Galli was obligated to give this humble class of healers, most of whom were illiterate, required a special kind of transitional teaching text. Anna Morandi and Giovanni Manzolini's three-dimensional visual guide to normative reproductive anatomy and manual obstetrical practice served to satisfy this need. The obstetrical waxes created by the couple provided a vivid and palpable synopsis of canonical texts and phenomena, Galli's work, and their own extensive knowledge of the reproductive body. From diverse primary and secondary sources, therefore, the couple created standard maps of the female anatomy comprehensible to even the rawest initiate.

Among Galli's more than 170 wax and clay obstetrical figures, only three have been firmly attributed to Morandi and Manzolini.[36] Each of these consecutively numbered figures represents the placenta with membranes and attached umbilical cord. The last also shows the cord's attachment to the uterus. With Hendrick van Deventer, Galli believed that the chief risk to a normal delivery was the attachment of the placenta to a spot other than the center of the fundus, the large upper end of the uterus.[37]

On 9 December 1746, Galli lectured on this theory to the Academy of Science in the Institute of Sciences with the aid of the first two of these figures sculpted specifically for the occasion. He later made use of the wax placentas when emphasizing in his household obstetrical lessons the importance of carefully examining the attachment of the placenta to the uterus. The accuracy and utility of these first figures prompted Galli to commission twenty more from the couple for a comprehensive introductory course in obstetrics for midwives. Primary and secondary sources suggest that these figures represented the female reproductive organs, as well as the changing uterus with developing fetus in progressive stages of pregnancy.[38] They thus combined current theory with empirical anatomy for practical pedagogical aims.

Morandi and Manzolini also worked closely with Pope Benedict XIV's appointed professor of surgery, Pier Paolo Molinelli (1702–1764), creating a number of exclusive wax figures for him.[39] Like their obstetrical waxes prepared for Galli, the waxworks made by the couple for Molinelli's use in his surgical courses added to his prestige and the reputation of his medical practice within and outside Bologna. The nineteenth-century biographer Michele Medici tells of one of the more celebrated of his models:

> A [pregnant] woman died six hours after delivery. According to Molinelli, no anatomist or obstetrician had ever brought to light the exact figure of a female uterus in the state in which it is found immediately after delivery, including [Josias] Weitbrecht and Johann Georg Roederer,[40] most praiseworthy authors at the time. Wishing therefore to correct this deficiency, he turned to Manzolini, requesting that he model in clay the uterus of that woman with all of the particularities by which the transitory state of that organ is accompanied: work executed with such precision and truth and of which Molinelli was so satisfied that he had a drawing made of it, which, with the uterus itself, he exhibited for viewing at his school in the introductory lecture to his course on surgical operations on cadavers. The uterus was copied and brought to even greater perfection by [Manzolini's] wife, Anna Morandi.[41]

For the paltry sum of thirty *scudi*, Molinelli also bought from the couple their series on the ear, considered "the most complicated" anatomical study in their burgeoning collection. According to the reigning president of the Institute of Sciences, Jacopo Beccari, these exacting figures "put to shame" the work of the renowned anatomist Antonio Maria Valsalva (1666–1723), who had pioneered the "new anatomy" of the ear while teaching at the University of Bologna.[42] Indeed, it is clear that Valsalva's study of the ear, immortalized in his *De aure humana tractatus* (1704), had been a chief inspiration for the couple's extensive series on the auditory organ and Morandi's famed colossal ear (fig. 38).

Morandi and Manzolini unmistakably aspired to affiliate themselves with and indeed to surpass anatomists who had special acclaim in Bologna. At the

FIGURE 38. Anna Morandi, colossal ear, wax. Courtesy of Museo di Palazzo Poggi, Università di Bologna.

time they were building their anatomical practice, Valsalva was perhaps the most venerated of Bologna's forefathers in anatomy. As Malpighi's brilliant student and Morgagni's teacher, Valsalva held a privileged place on Bologna's genealogical tree of anatomical genius, a tree on which Anna Morandi and Giovanni Manzolini clearly hoped to claim their own branch. Indeed, the anatomist-modelers may have had an even more ambitious desire to attach their names to the broad ancestral line of innovators across the peninsula in the *new* anatomy, beginning with Galileo and his iatromechanical conception of the ear as an acoustic machine. Giovanni Alfonso Borelli (1608–1679) extended Galileo's mechanical theories of anatomy and taught these to Marcello Malpighi, the father of microscopic anatomy, who in turn passed his discoveries of infinitesimal pulleys, levers, and other mechanisms of which he believed the human body to be composed to his student Valsalva. Valsalva undertook with his own gifted student, Giovan Battista Morgagni, a sixteen-year empirical study of more than one thousand human heads to discern the mechanical structures that formed and operated within the ear.[43] Valsalva's study of the anatomy and physiology of the ear, which he interpreted much like Galileo as an intricate machine designed to decipher sounds, culminated in 1704 with the publication of his masterwork *De aure humana*. This book would serve as a point of departure and an explicit foil for the theoretical and visual studies of the ear by Morandi and Manzolini. Not only were they determined to surpass Valsalva in the precision of their experimental studies of the ear, among the most mysterious and complicated organs in the body, the couple sought to amplify their sphere of expertise, appropriating to themselves authority over not only the ear, but the four other sense organs as well—the eye, the mouth, the nose, and the hand. Estiology (the study of the sense organs) would in fact continue to be, after Manzolini's death, a central province of Morandi's anatomical expertise. As will be seen in chapter 5, which explores her studies of the eye (the organ of sight) and hand (the organ of touch), Morandi would also exceed the laws of mechanism in her conceptualization of the anatomy of perception.

Two tracts on the anatomy of the ear authored by Manzolini are among the most important records of the couple's scientific pursuits, their method, and their ambition to establish themselves within the elite Bolognese academic community as expert *professors* of anatomy, especially estiology, and to thereby supersede Valsalva's legacy. Although Morandi's name did not appear as coauthor of the tracts, she doubtless collaborated in the dissections and the analysis of anatomical texts and images at their core. Given that her notes and wax anatomical figures would reiterate and in some cases extend the principal theses proffered in these tracts, it is quite plausible that she also helped to write them.

The first tract, written on 9 March 1749 and read two years later on 27 March 1751 at an assembly of the Benedictine Academy of Science within the Institute of Sciences, aimed to decide once and for all the age-old question of the origins of mutism in the congenitally deaf. Manzolini recounts that the opportunity to resolve the issue arose at the conclusion of the recently held Public Anatomy, when he learned that the anatomist Domenico Galeazzi had "a surplus of parts of Cadavers, including the head and neck of a thirty-five-year old subject, who had been born and lived to the end of his life without hearing or speech."[44] Manzolini immediately requested and was granted by Galeazzi this severed head for the purposes of dissection. In the couple's home studio, Manzolini, in presumed collaboration with Morandi, undertook a comprehensive examination of the "nerve and muscle structures that serve elocution," finding that these were without any injury or defect that might impair speech.[45] However, the subsequent examination of the auditory organ instead showed that, among other severe abnormalities, the cochleae in both ears were essentially empty cavities lacking the normal spiral structure necessary for hearing. Manzolini thus concludes that the subject "was mute for having been born deaf,"[46] and was able to postulate that mutism is a standard consequence of congenital deafness.

The second treatise was written on 7 February 1750 and read two months later at another assembly of the Benedettini by the reigning institute president, Jacopo Beccari.[47] Indeed, Beccari likely invited Manzolini to submit his tracts on the ear to the academy as a way for the beleaguered artist to bolster his reputation as an anatomist in the aftermath of his well-known dispute with Ercole Lelli. Divided into two parts, the tract first sets forth an original theory on the form, topography, and function of the "auditory nerve," by which Manzolini was actually referring to the facial and vestibulocochlear nerves. In the second part of the tract, in answer to a formal request from the Academy of Science, and most probably President Beccari, Manzolini identifies numerous inaccuracies in the illustrations of the anatomy of the ear in Antonio Valsalva's *De aure humana*.

Giovanni Manzolini opens this catalogue of corrections in a tone belittling of his famous precursor. Accentuating his superior proficiency in the *new* anatomy and his just impatience with Valsava's scientific imprecision, Manzolini observes: "In Table One, Figure Three, [Valsalva] assigns to the external Ear, three posterior muscles, marked CCC of which, in truth, no more than one is usually found, rarely two, but without the help of a knife to produce it, never three." Although Manzolini refrains in the rest of his catalogue from suggesting that his forerunner was so flagrantly inept as to have illustrated anatomy that was a fabrication of his own clumsy scalpel,

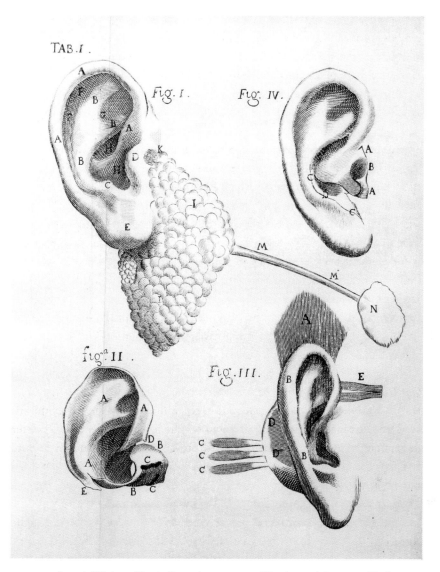

FIGURE 39. Antonio Valsalva, table 1, in *De aure humana tractatus* (Utrecht, 1707). Courtesy of Becker Medical Library, Washington University School of Medicine.

he indicates in each correction Valsalva's failure to accurately illustrate "the truth" revealed in dissection (fig. 39). He states, for example, that in the second figure of Table Eight, "five openings are illustrated labeled *g* at the exterior of the Vestibule [labeled] *f*, the use of which [Valsalva] claims is for an opening for the soft portion of the auditory nerve at the interior of the labyrinth. Yet, no more than one [opening] is actually found" (fig. 40). Manzolini likewise found the sixth figure of the same table to grossly misrepresent the shape and nerves of the seventh pair called the auditory nerves," written in all likelihood years earlier:

> the auditory nerve is divided into two portions, one hard and the other soft that are united together at the point of the bony transverse at which place they re-unite with the soft portion filling the large opening [*fossetta grande*] and meeting up with the base of the snail [*lumaca*, by which he means the cochlea, the standard anatomical term later used by Morandi] except for some small threads that extend along the base to the Vestibule. The hard portion thus separated from the soft by means of said bony transverse, extends to the small opening [*fossetta piccola*] at which point it forks entering by means of this bifurcation two particular openings . . . the one being the origin of a particular canal called the fallopian aqueduct [facial canal] and through this the larger soft branch of the hard portion runs afterward providing other areas with its various ramifications.[54] The other opening at the Vestibule receives the smaller thread of this portion that has always been thought to terminate at the circular base of this opening. Yet, one consistently observes that it proceeds to the entrance of the Vestibule where and in such manner that it becomes the nerve of a fascicle of meaty fibers.[55]

Manzolini's description is nearly as tortuous as the complex anatomy of the ear itself. Yet in fairness, he was attempting to articulate the most exacting description of the auditory nerve yet produced. Both he and Anna Morandi, in fact, had a remarkably accurate understanding of the auditory organ based on scrupulous dissections, yet, as Luigi Galvani would zealously commemorate in his eulogy of Morandi, she set the highest standard of accuracy yet seen in her wax representations of the near invisible intricacies of the anatomy of the inner ear:

> Observe, finally, within the ear (overlooking other figures) not only that most subtle nervous ramification that constitutes the chorda tympani, but also the ramifications of the *hard* nerve that extend to the internal muscle of the malleus and to the stapes. As minute as these are, almost to the point of escaping our sight, you nevertheless see them arranged inside the tympanic cavity, a cavity so small that you could not even insert your finger. See them loose, extended and attached to said muscles![56]

In contrast to her husband, Morandi would never explicitly contradict Antonio Valsalva in her study of the ear, although she did not hesitate to challenge the theories of other master anatomists. By the time she began to conduct solo anatomical studies and demonstrations after her husband's death, it appears no longer to have been necessary for her to prove her proficiency in the anatomy of the ear at the expense of the author of *De aure humana*. Instead, in her notes on the ear she disputes only one contemporary and relatively minor figure, Johannes Augustus Quirinus Rivinus, on his theory of an "incisura tympanica," or hiatus in the tympanic membrane.[57] Rivinus announced his theory in his treatise *De auditiis vitis* published in Leipzig in 1717; Morandi most probably came to know of it in Albrecht von Haller's 1749 republication of the tract in his *Disputationes anatomicae selectae* (Göttingen, 1746–51, vol. 4). With striking assertiveness, Morandi argues against Rivinus's thesis in a lengthy digression on a popular experiment to force smoke into the ear canal:

> I am pleased to report here that I can not accept the sentiment of Rivinus, who would like to find in the middle of the Tympanic Membrane an opening that in truth is not there, perhaps persuaded of his opinion, as with certain others who adhere to it, by the transmission of Smoke through the Ear by some who today smoke tobacco. . . . In my view, however, this does not happen through the opening that Rivinus is wont to find in the middle of the Tympanic Membrane, which is in no way to be found there. Rather, because the external Muscle contracts after being irritated by the smoke that is forcibly introduced and that swirls inside the Tympanic Cavity, it moves, albeit inadvertently, or lessens its adhesion to the Tympanic Membrane; and because of this contraction leaves a passage open for the smoke that was introduced into the Tympanic Cavity via the Eustachian Tube.[58]

Morandi's focused critique of this theory served to highlight her familiarity with current anatomical texts and ideas. At the same time, Valsalva and indeed Haller were indirectly implicated in her analysis. Haller, who had authorized Rivinus's theory in his republication of *De auditiis vitis*, was a lightning rod in Bologna at the time for his theories of irritability. And Rivinus's hypothesis also involved the so-called Valsalva maneuver, the method invented by Valsalva of testing the openness of the eustachian tube by having his patients forcibly exhale while keeping their nose and mouth closed. Morandi's critical assessment of Rivinus thus linked her, albeit loosely, with noted contemporary authorities in anatomical science. She thereby staked her claim to a privileged place on the genealogical tree of leading anatomists.

High-Status Commissions and Grand Tour Demonstrations

Beyond their native city, Anna Morandi and Giovanni Manzolini received prestigious commissions to sculpt anatomical wax figures from King Charles Emanuel III of Sardinia, King Charles of Naples, and the Venetian procurator Alvise IV Mocenigo, who became doge in 1763.[59] According to contemporary biographers who had a firsthand acquaintance with the couple, as well as their own direct testimony in letters,[60] their reputation for exquisite anatomical models extended to the premier courts and academies across Europe. They created models for King Augustus III of Poland, and their series of the sense

However, it is Morandi's epic bodily fortitude in the practice of this grueling science that most sets her apart in Bianchi's estimation. In contrast to the better part of her sex, she is able to stomach an elbow-deep familiarity with the viscera of the dead; yet, it is her maternity that proves her extraordinary stamina to Bianchi and grants her exceptional knowledge of the body. For Bianchi, Morandi is quite literally Bologna's *Madre degli Studi* of human anatomy:

> Not to be repulsed by the practice of anatomy is a very rare quality in a woman, and she is still young enough to bear children frequently. Yet, even after delivering her young she does not neglect her dissections or her wax preparations, which she has sent to the Kings of Poland, Naples, Sardinia and to other dignitaries who have commissioned them, prompting one to say that this is the first woman who has achieved so much with respect to the field of practical anatomy. Moreover, through birthing her offspring and anatomizing cadavers, she achieves in both her synthetic and her analytic method a great understanding of the human species.[71]

Bianchi praises Morandi for applying to anatomical science an incomparable version of the bipartite method of inquiry developed by the Greeks and newly conceived by the lights of experimental science and philosophy.[72] The two prongs of the one method are analysis, whose particular conclusions are drawn from observation and experiment, and synthesis, the systematic and composite verification of those same conclusions. Galileo, to whom Bianchi looked as supreme exponent of the "new science," placed the definition of this scientific method on the tongue of his own mouthpiece, Salviati, in his *Dialogue concerning Two Chief World Systems*. In answer to the Aristotelian foil Simplicio's avowal that *the Philosopher* proceeded first "*a priori*, . . . by means of natural, evident, and clear principles," and then supported his conclusions "*a posteriori*, by the senses and by the traditions of the ancients," Salviati reverses the approach in keeping with the *new* scientific method: "I think it certain that he first obtained it by means of the senses, experiments, and observations, to assure himself as much as possible of his conclusions. Afterwards he sought means to make them demonstrable."[73] For Bianchi, Anna Morandi's *analysis* is thus her hands-on dissection, discovery, and identification of the parts and structures of the body. Unlike conventional (male) practitioners, however, Morandi is capable of a matchless synthetic procedure of final demonstration. She verifies her analytic results by means of the revelation of her own living viscera turned inside out through childbirth. Bianchi affixes to the body of the woman anatomist herself what Katharine Park has described as the "special, symbolic weight [of the uterus]" that had long been *the* sign in anatomical illustration of "the body's hidden interior."[74] Morandi's knowledge and instruction of anatomy is thus uniquely informed by the composite test of generation and by the unbolting of the subterranean vaults of her own body. Simply put, she ascertains the structure of the body by taking it literally to pieces through dissection—and by putting it together, forming a whole body through maternity. And unlike her male counterparts, she can look fearlessly on her interior, and she will not turn to stone.[75]

Morandi's Considered Response

Two letters written to Bianchi by Anna Morandi in April and May of 1755[76] leeched the hyperbole from his depiction of her as the Mother of All Anatomists. In place of her maternity, Morandi underscores in a tone of cool confidence the esteem she and her husband enjoyed for their expertise in anatomy among leading professors of medicine and in centers of learning in Italy and across Europe. Morandi's letters were, in fact, written as a rebuttal to a letter published by Bianchi in which, without naming them, he called into question her and her husband's anatomical competence.[77] The controversy is significant for the extensive information it provides about Anna Morandi's and Giovanni Manzolini's actual engagement with the community of scientists and medical practitioners. It merits careful examination also for what it reveals about the way in which Morandi represented her anatomical practice with her husband, and their authority among leading members of the scientific community.[78]

The Case of the Contino Pilostri

Behind the epistolary exchange between Giovanni Bianchi and Morandi was a contentious cause célèbre among medical practitioners across a northeastern swath of Italy, from Cesena to Rimini and Padua to Bologna. Known as the case of the Contino, or child count, the controversy erupted over the pathology of a bone fragment of nine-year-old Giambattista Pilostri, to whom Bianchi had ministered in 1749 at the end of the child's life and autopsied after death.[79] The Pilostri case became widely known in 1750, when Bianchi began citing it in repeated publications aimed at advancing his theory on the cerebellum's and the cerebrum's distinct functions and influences over the lower body.[80] He claimed that in his autopsy of the boy's brain, he had found a lesion in the right lobe of the cerebellum that had caused paralysis on the right side of the child's body, and not the opposite side, as would have occurred with a similar malignancy in the cerebrum. Bianchi sought approval for his "Storia Medica" [Medical Case History] from noted scholars, including the revered Giovanni Battista Morgagni at Padua.[81]

The battle over the case was not fully joined, however, until 1755, six years after the death of the child, when Carlo Serra, a physician from Cesena and a bitter rival of Bianchi's who had also tended to the dying boy and assisted in the autopsy, published his own account of the case. In it he denounces Bianchi for medical incompetence and misdiagnosis of the boy's fatal illness. Serra claims that the boy had been afflicted instead by "fluxions in the ears" (an excessive flow of poisonous humors). Not content to clash across the printed page, however, Serra took his dispute with Bianchi on the road and sought to discredit him in prime centers for the study of medicine, including Padua and Bologna. According to several articles on the controversy published by Giovanni Lami in the *Novelle letterarie*, during the autopsy of the boy Serra had secretly removed a fragment of the temporal bone. It was covered, he claimed, in cavities that confirmed the boy's fatal case of ear fluxions.[82] Serra brought the pocked bone fragment with him to Padua and Bologna to prove his theory and discredit Bianchi.

It was to this public attack by Serra that Bianchi responded in his published letter of 1755. In it he denounces Serra together with the two unnamed Bolognese anatomists who corroborated Serra's judgment on the cause of irregular fissures in the young nobleman's skeletal remains. Bianchi writes with mocking praise of the "two most skilled Bolognese professors" (due Bravissimi Professori Bolognesi), who made a fine copper engraving but a wholly incompetent anatomical analysis of the boy's temporal bone:

> These most skilled Bolognese Professors [Morandi and Manzolini] . . . may very well have been able to etch that [skull] fragment, according to their art, but they were not capable of truly judging if the bone was diseased, that is rotten, or wormeaten, since a bone taken from a Cadaver, which has not been cleaned, and which was not boiled in water in order to cleanse it well of all flesh . . . deteriorates and reeks, and to the inexperienced can seem diseased, which those valiant Professors should have known to whom [Serra] said he went with this portion of the bone, which only now, after six years, he brings before the Public. But without saying who these valiant Professors are, all can well doubt with me their judgment and their valor.[83]

In her epistolary response to Bianchi, Anna Morandi offers a very different account of the couple's leading role in the strange chain of events that followed Carlo Serra's arrival in Bologna flaunting the little count's pocked skull fragment. Morandi reports that after taking the piece of bone for evaluation to Dr. Morgagni in Padua, Serra "brought it to Bologna and had it examined by, among others, signori Galeazzi, Laghi, Beccari, Galli, Molinelli, and Bibiena," the city's premier medical scholars.[84] The illustrious Bolognese doctors all came to the same conclusion, "that the bone was pocked." To verify their judgment, Drs. Beccari and Molinelli urged their colleagues Domenico Galeazzi and Tommaso Laghi to bring Dr. Serra secretly to the Morandi-Manzolini studio so that the couple could determine "if, in that fragment of bone, cleaned and polished through valid maceration, cavities could be detected." The couple was also asked to make an "exact design" of the skull segment in the form of a bronze engraving. While Morandi explains that this was well outside their regular practice as wax modelers, she and her husband "agreed to undertake the task with some not unpromising results for our trouble."[85]

In a few sharp phrases, Morandi roundly refutes Giovanni Bianchi's allegation that she and her husband were incapable of accurately evaluating the skull's fissures:

> As regards the [bone's] being pitted, it was plainly manifest because of its very obvious pockmarks, in the exact manner and no other as demonstrated in the copper engraving. We could thus do no less than unite ourselves to the sentiment of so many excellent Men, who had affirmed the same. Nor did it require much investigation to determine so evident a fact by those, who day and night are occupied according their profession in the execution of anatomical sculptures, as we are.[86]

Morandi bolsters her authority, moreover, by underscoring the international celebrity and patronage the couple enjoyed for their mastery in anatomy: "And it is known not only to your most Illustrious and Excellent Lordship, but to all of Europe, that many Academies acquire our [models] and pay us the honor of utilizing them in the instruction of students of anatomy." She undoubtedly recalls here Bianchi's own praise for the Lady Anatomist's continentwide prominence in the letter he had published the previous year in the *Novelle letterarie*. Morandi finally concludes her letter with a circuitous but unmistakable admonition of the good doctor: "While to assail [*toccare*] us in your letter might advance your most Illustrious and Excellent Lordship . . . we will not hesitate in due time in our own writings to seek a remedy."[87] Notwithstanding the requisite courtesy of her reply, Morandi summarily invalidates both Bianchi's criticism of the couple and his previous portrait lauding her *maternal* science.

Morandi's second letter of 24 May 1755 reiterates the same perfunctory courtesies but more concretely buttresses previous arguments in defense of the couple's expertise in anatomical science. She claims that while she and her husband would never fail to esteem properly the learned doctor, yet

> neither are we in any way disturbed that your most Illustrious and Excellent Lordship has said that, because we are not surgeons, we are incapable of evaluating the cavities in a bone, which goes to show that Your Most Illustrious and Excellent Lordship is not well informed that we have over many years dissected hundreds and hundreds of cadavers and kept aside the bones of these for our

preparations. Because of this, we have had a large field in which to discover even too many of those cavities in broken bones and others unbroken.[88]

Anna Morandi's key defense rested significantly on the vast extent of the couple's experience at the dissection table. She pitted their anatomical experience and proficiency against any practitioner in the field: surgeon, anatomist, pathologist, or physician. However, she also went further in this letter to repair the strained relations with the doctor. She conceded Bianchi's claim that he had found no cavities in Pilostri's temporal bone when he conducted his autopsy six years earlier, and so concluded that the bone fragment from Dr. Carlo Serra that she and her husband had analyzed and reproduced in a copper engraving was not therefore that of the boy count.[89] Morandi struck an astute balance in her correspondence with Bianchi between firm self-justification and political tact. In the two letters she wrote to the physician, hers was clearly not the voice of an improviser either in the anatomy theater or the theater of public opinion. She spoke from the position of authority in which she was well ensconced by 1755 as an expert anatomist and anatomical demonstrator.

Ample testimony of native and Grand Tourists to Bologna and of the woman anatomist herself thus contradict pervasive representations of Anna Morandi as anatomical extemporizer apprenticed to her husband until his sudden death in 1755. As this chapter has sought to show, she instead emerges from primary texts and correspondence written during more than a decade before Giovanni Manzolini's demise as a recognized actor, indeed a *novatrice*, or innovator, in the field of anatomical science in and beyond Bologna. In her household classroom she taught the stratification and physiology of the human body in two essential and affiliate ways: through the conventional means of the dissection with scalpel and forceps of human cadavers, and by means of her meticulous and incorruptible anatomical facsimiles. She professed not merely the empirical lessons of the corpse, but the extra-experiential knowledge of the inner workings of the living body, its intricacy and the discrete and collective utility of its components to which the corpse could only allude. Although she did not reiterate in her studiously prepared anatomy demonstrations either the spectacle of the *improvvisatrice*'s performance or Laura Bassi's staged presence at the Carnival dissection, she was, of course, spectacular in her own right. As we have seen, her gender was a powerful, at times overwhelming, draw for tourists to her anatomy lessons. She was an anatomist, but more important, she was the Lady Anatomist, who skillfully cut open cadavers and modeled exquisitely in wax, all while her melancholic husband seemingly disappeared behind the widening shadow she cast. That she was well aware of her image and sought to influence her public representation in terms of gender, class, and professional attributes is best demonstrated by her wax self-portrait, to which we now turn.

4

The Lady Anatomist

Women should not be excluded from the study of the sciences, since their spirits are more elevated and they are not inferior to men in terms of the greatest virtues.

GIUSEPPA ELEONORA BARBAPICCOLA. "La traduttrice a' lettori"[1]

The life-size wax self-portrait of the eighteenth-century Bolognese anatomist and artist Anna Morandi represents the most evocative narrative of her enigmatic lifework, and one with which all other accounts must necessarily contend (fig. 41). In this visual autobiography, she depicts herself outfitted with sharp-edged dissecting instruments—a scalpel and forceps (now lost), as well as a whirl of feminine finery—peach taffeta, gray lace ruffles, festooned sleeves, and lavish faux jewels. In diametric opposition to the show of strung pearls, flowing hair, languid, voluptuous gaze, and supine pose that would symbolize the wax Venus's surrender to the authority of the Florentine anatomist Felice Fontana at the century's close, this disquieting, even defiant, pastiche of conventionally alien signs silently proclaims Morandi's distinction in the field of anatomical science (fig. 42).

She is the Lady Anatomist, a spectacular paradox manifested by the distance between her flamboyant form and punctilious function. Elegantly arrayed and casting her serene gaze outward at her viewers, her hands prepare to penetrate the human brain mushrooming from the hair-covered skull before her. In place of the customary plain dress of her social class, her attire denotes high social status and is unmistakably meant to inflate her rank.[2] More subtly, however, her refined clothing serves to countervail the ignoble task of anato-

FIGURE 41. Anna Morandi, self-portrait, wax. Courtesy of Museo di Palazzo Poggi, Università di Bologna.

FIGURE 42. Clemente Susini, anatomical Venus, wax; detail. Courtesy of the Specola Museum of Natural History, Florence.

mizing the contents of the open skull as well as to underscore her sex.[3] As Mary Sheriff has shown, artists often featured in their self-portraits ennobling signs of aristocratic dress to "disassociate the manual labor" of the art and, in this case, of the fouler work of anatomical science as well. Yet this marriage of dichotomous signs—ornate jewelry and scalpel; sumptuous raiment and anatomized body parts—is not fully joined. The seat of knowledge that lies quite literally in her feminine grasp is a clear provocation to those who would doubt women's intellectual authority (fig. 43).

Juxtaposed to Morandi's image and extending its connotations is the companion wax bust she sculpted of her late husband, Giovanni Manzolini, that shows him dissecting a human heart, the traditional seat of passions (fig. 44). He exhibits an outward, slightly downcast gaze and dissecting hands—his right hand holds forceps while the left rests directly on an extricated and unfurled heart. The historian Giovanna Perini has observed that Manzolini's

FIGURE 43. Anna Morandi, self-portrait (fig. 41), wax; detail. Courtesy of Museo di Palazzo Poggi, Università di Bologna.

FIGURE 44. Anna Morandi, portrait of Giovanni Manzolini, wax. Courtesy of Museo di Palazzo Poggi, Università di Bologna.

severe black tunic, jaundiced complexion, conspicuous beard stubble, and far-away look render him a melancholy figure.[4] Although Morandi undoubtedly sought to immortalize her husband's contributions to anatomical science as well as to honor their professional partnership while he was alive, her portrait of him presents a studied contrast to her own vibrant and elegant aspect and acts to heighten her distinction.

For au-courant Bolognese, the depiction of Manzolini anatomizing a heart would also have been a subtle but clear allusion to the furious midcentury debate on the question of "irritability" that gripped the city's *accademici*. Key defenders in Bologna of Albrecht von Haller's controversial theory that an inherent force (*vis insita*) causes muscle contraction, including Leopoldo Marco Antonio Caldani, Pier Paolo Molinelli, and Eraclito Manfredi, supporters all of Anna Morandi and her anatomical practice,[5] viewed the heart as *the* organ of irritability, since it could be seen to beat for some time even after death. A veritable mass slaughter of cats and dogs, in fact, took place in Bologna's anatomy laboratories to reveal that distinctive power "hidden in the fabric of the heart."[6] The concept of irritability and the academic quarrel it provoked in Bologna arose, notably, after the death of Morandi's husband. By portraying him as the anatomical authority of the heart, as opposed to the ear, for which he was most famous, she amplifies the currency within the local academic arena of his anatomical practice and, of course, her own.

With her self-portrait and the suggestive and expedient image she created of her late husband, Morandi constructed a partial narrative about her life and work. She glorified her distinction as a woman anatomist and the import and contemporary bearing of the anatomical subjects in which she was expert while only obliquely alluding to her role as an artist. The vibrant and meticulous image she sculpted of herself implicitly affirmed her skill as wax sculptor and anatomical designer. But this was mere subtext to the principal tale of her conquest of human anatomy, the viscera of the "new science" that she had wrested from men's exclusive hold.

Morandi's self-portrait recalled representations of (male) anatomists and the public dissection scene conventional to the title pages of sixteenth- to eighteenth-century anatomical texts. That she was well acquainted with these illustrations is confirmed both by the extensive catalogue of anatomy books in her private library—which, as we have seen, included, among others, the atlases of Vesalius, Valverde, Bartholin, and Cowper—and by the frequent references and corrections she made in her scientific writings to these and other authors.[7] Andrea Carlino's deft explication of the dissection scene in early modern anatomical texts provides a theoretical frame for locating Morandi's self-portrait within this iconographic tradition.[8]

According to Carlino, title pages of early anatomical texts depict dissection scenes consistent with the quodlibetarian, or scholastic, model of anatomical study.[9] Three actors separated by space and status are shown performing the anatomy lesson: the *lector* sits on the pulpit above the dissecting table, reading from the authorized Galenic text; the *ostensor* stands (most often with a pointer) next to the cadaver and indicates the parts of the body under discussion; and the lowly *sector*, a barber-surgeon, directly handles and dissects the cadaver (fig. 45). The *lector* typically looks away from the body, outward at the reader or at the anatomical text he is consulting. The upward gaze represents the sublimation of the dissection by means of scientific theory and authorized texts. The *ostensor* is level with the body and looks upon it, but at the distance of his pointer, while the *sector* hovers close to the cadaver he touches and rends. The title page of Andreas Vesalius's *De humani corporis fabrica libri septem* (1543) signifies, according to Carlino, a shift in both the practice and the representation of the anatomical lesson (fig. 46).

No longer remote from the anatomical subject, "Vesalius is portrayed with his right hand thrust in a woman's abdomen and his left raised as if to emphasize the words accompanying the demonstration."[10] The image serves to extend Vesalius's vigorous critique within the text of the exclusively theoretical and authorized model of anatomical inquiry, which he supplants with his own example of direct observation and dissection of the cadaver. In the background, each of three audience members holds an anatomical text, two open and one closed, demonstrating the subordination of received knowledge to firsthand experience of the body. However, Vesalius, like the lector of old, looks away from his subject and his surrounding audience, toward his readers. As Carlino notes, he is not shown consulting a text in the manner of the traditional lector, but instead appears to be writing his own book of anatomy based on immediate observation, suggested by the quill, ink, and paper on the dissecting table.[11]

Anna Morandi's self-portrait closely follows the Vesalian iconographic model through its integration of the academic, didactic, and practical functions of the anatomist. As in the case of Vesalius's image, her self-portrait maintains what Carlino has called the distance between the space of dissection and the academic space through an epistemological interval between her active hands and her contemplative gaze. Like the traditional lector, Morandi looks thoughtfully upward, away from the anatomical subject, but as a *new* anatomist she performs the dissection herself. She unifies in her own person the practical and the theoretical features of modern scientific practice, what Roy Porter eloquently deems "the Baconian union of hand and head, mental and manual labor."[12] In contrast to the Vesalian model, however, where the

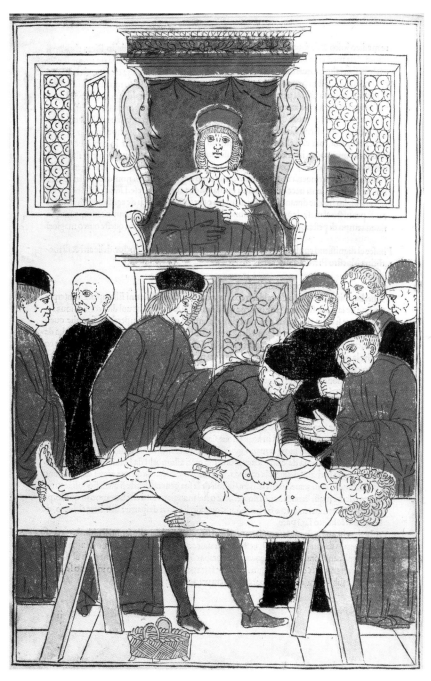

FIGURE 45. Johannes de Ketham (attributed), title page of *Fasciculus de Medicinae* (Venice, 1493). The Metropolitan Museum of Art, New York. © The Metropolitan Museum of Art / Art Resource, NY. Photo: Metropolitan Museum of Art / Art Resource, NY. Used by permission.

FIGURE 46. Andreas Vesalius, title page of *De humani corporis fabrica libri septem* (1543).

book once held by the lector recedes into the background but is nevertheless present, there is no overt reference to an outside textual authority. Morandi alone embodies the authority of anatomical science in this dissection scene. Her self-portrait acts, in fact, as title page and theoretical preface to her vast, three-dimensional, wax anatomical atlas that spanned the room and lined the walls of her household studio until it was moved to a designated museum within the Institute of Sciences two years after her death, in 1776. The boisterous throng of university doctors, students, foreigners, and curious onlookers conventional to both the actual Public Anatomy and the illustrated dissection scene has been replaced by us, Morandi's living pupils and admirers, on whom she turns her direct gaze. Gone, too, are such customary memento mori as the animated skeleton and écorchés that loom menacingly and often in colossal proportions on the opening pages of early modern anatomical atlases. Death is figured here realistically by the skull, to which the matted human hair of the implied organ donor still clings. Indeed, the practical objects and instruments of scientific knowledge—scalpel and forceps—have replaced such standard metaphysical signs as the skeleton and the cherub, indicating an epistemological as well as an iconographic shift.

Morandi's representation of herself is certainly spectacular and indeed spectacularly enhanced as regards her social status and sartorial refinement, especially in contrast to the way she actually dressed when cutting into a brain in her anatomical dissection laboratory. As we have seen, her anatomical notebook confirms that she was fully cowled against the reeking and leaking of the "fresh parts" she dissected.[13]

The most striking change manifested in Morandi's self-portrait, however, is its provocative inversion of gender roles. Vesalius's manual exploration of his female subject's abdominal cavity had been supplanted by the Lady Anatomist's dissection, at the distance of her surgical instruments, of a human brain. The presence of a woman in an illustrated anatomical lesson came nearly always in the form of the cadaver. Rarely was she a member of the viewing public, and never was she the professor-anatomist. Morandi provocatively asserts her expertise in what was for eighteenth-century Enlightenment thinkers that most critical realm of the body. Her patron, Pope Benedict XIV, described the brain as the "most secure seat and domicile of the soul," and the site of the imagination, which was in turn the "book of the brain," wherein are written all "intellectual notions as well as the images of perceptible objects collected by the senses."[14] The authority she ascribes to herself in her wax facsimile tacitly extends not only to the brain's anatomy and its myriad dependent parts, but also to the crucial functions it performs and governs, namely cognition and perception. Morandi was most noted for her extensive studies in word and wax of the sense organs, whose perceptive powers may be seen to stem from and flow to that concave mass of nervous tissue. The unorthodox position of power and intellectual authority in this scene of dissection that Morandi created was prelude to the unorthodoxy of her oeuvre.

The brain she puts under her knife in the self-portrait, along with the principal themes of her oeuvre, the sense organs (the eye, hand, mouth, nose, and ear), the male reproductive system and genitalia, and the gravid uterus, signify a provocative sphere of expertise at distinct odds with contemporary notions about women's inferior nature and restricted intellectual and moral purview. Morandi's self-representation and the parts of the body on which she focuses her gaze are thus central to what was novel about her work. She figures herself

she might need" from the Hospital of Santa Maria della Morte. And these body parts, the directive makes clear, must be sent to her "immediately and in wholesome condition."[23]

The flurry of honors, appointments, and financial assistance granted Anna Morandi shortly after her husband's death is clear proof of official sanction of her work. However, despite appearances to the contrary, that patronage was also unmistakably lackluster. Bolognese leaders acted on fear of the legitimate threat that, without formal recognition and recompense from the state, Morandi would move her practice and famous waxes to another more solicitous city. Documents of the Senate, the pope, and university administrators all refer to that threat. Yet in the end, their fear was not so deep as to prompt them to grant her more than an honorary place in the Clementina and a token honorarium of 300 *lire* per annum. Ercole Lelli, of course, had enjoyed membership in the institute and the posts of Custodian of Museums and Coiner, as well as regular membership in the Academy of Art from 1742, where he held the powerful offices of Director of Figure Drawing, Vice Prince, and the highest office of Prince in 1747 and 1754, the year before Morandi's honorary induction. Compared to Lelli's annual stipend at the time of 1,200 *lire* and Laura Bassi's 760 *lire* (increased in 1759 to 1,000 *lire*),[24] Morandi's was indeed a humble reward. As her family circumstances would immediately prove, it did not amount to a living wage.

On the third of October 1756, less than a year and a half after her husband's death, Anna Morandi surrendered to the Orphanage of San Bartolomeo di Reno her eleven-and-a-half-year-old son, Giuseppe Maria Gioacchino, and all parental rights to him. The transfer actually had been prearranged on 16 November 1755, only six months after Giovanni Manzolini's death, when Morandi's dire financial circumstances were already plain. Giuseppe remained a ward of the orphanage for two and a half years until, on 21 February 1758, records show that he was "removed from the orphanage in the presence of the Monsignor General Vicar," to whom the boy was then consigned in order to be brought to the noble Solimei family, of which he would become the official heir. The Solimei had no blood descendents to whom to pass on their fortune and their name.[25] From the moment of his release from the orphanage, the boy was no longer Giuseppe Manzolini but became instead the young nobleman Giuseppe Solimei.[26]

Building Her Practice and Her Fame

For the nine years following her husband's death until 1764, when she suffered a prolonged, near-fatal illness and was forced from her home laboratory to convalesce in the countryside, Morandi relentlessly strove to advance her anatomical practice and her international fame. Hers was a campaign of self-promotion and self-preservation. It was during this period of intense labor at the dissection table and before a steady flow of international patrons of her household anatomy lectures and wax models that she distinguished herself as the best of the three Bolognese wax modelers with her expert knowledge of human anatomy and the accuracy, durability, and thematic intrigue of her models.[27]

During these nine years, Morandi completed her two-part oeuvre, the series of wax figures that comprised her three-dimensional anatomical atlas and the "magnificent book"[28] of companion notes that explained them. As has been said, most important among these works were the series on the sense organs, which she duplicated for numerous patrons, and the series on the male reproductive system, whose notable value would be confirmed upon its sale in 1769 for ten times the price of any other part of her oeuvre.[29] Her anatomical studies extended well beyond these two subjects, however, and comprised the skeletal system, expansively represented by actual skeletons from fetal stage to adulthood and by numerous display tables of discrete bones; and figures of the larynx, the pharynx, and the myology of the face and torso. She also exhibited in her studio and analyzed in writing and for visitors to her home the series of the forearm and the leg and foot that she had made together with her husband.[30] She put on display but did not annotate the figures she made with Manzolini of the heart and lungs. It seems that she created additional figures in wax of the female reproductive system and gravid uterus for use by Giovanni Antonio Galli, although records for these commissions are lacking.

The waxworks she conceived, as visitors repeatedly noted, were startling in their scientific accuracy and their visual grace. Of the three Bolognese *ceroplasti*, in fact, she came closest to achieving a perfect balance between the two. While her manner of representing the anatomical body was, for obvious reasons, a close match to that expressed in the figures she and her husband created together, the works she alone authored possessed a distinct lightness and mobility. See for example the bulky, inert figures of the respiratory apparatus and the heart that Manzolini and Morandi had made together (fig. 47). The couple's elegant, ideally proportioned, muscular leg, unmoving and fixed compliantly on the display table within its frame of waxen cloth, likewise seems the epitome of noble stillness (fig. 48) when compared to Morandi's forcefully protruding tongues, grasping hands, and gaping eyes that seem to transgress the confines of their display tables and enter the space of the spectator (figs. 49–50).

Across Italy and Europe, the academic world took notice of the genius and novelty of Anna Morandi's work at the intersection of science and art.

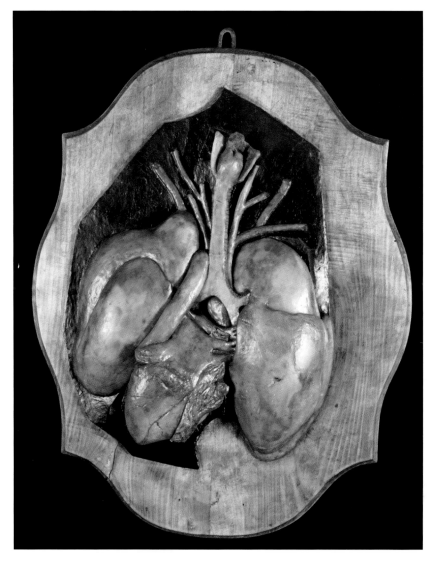

FIGURE 47. Giovanni Manzolini and Anna Morandi, lungs, heart, larynx, and pharynx, wax. Courtesy of Museo di Palazzo Poggi, Università di Bologna.

Her contemporary biographers, Luigi Crespi, Marcello Oretti, and Giovanni Fantuzzi, all of whom had a personal acquaintance with her, attested to the official invitations and accolades she received at this time from cities and academies within and outside Italy.[31] They asserted and records confirm that she was inducted into the prestigious Florentine Academy of Design, becoming a member on 11 February 1761.[32] Although no documents have surfaced to prove that she was sent a "blank check" from the city of Milan to move her practice there, as her biographers claimed, it was surely feasible.

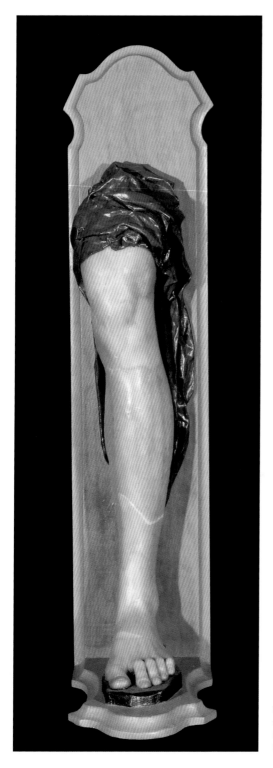

FIGURE 48. Giovanni Manzolini
and Anna Morandi, male leg, wax.
Courtesy of Museo di Palazzo Poggi,
Università di Bologna.

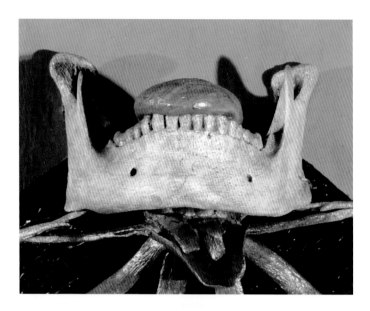

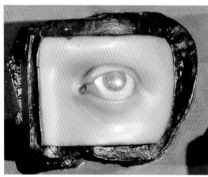

FIGURE 49–50. Anna Morandi, tongue and jaw, and eye, wax and bone. Courtesy of Museo di Palazzo Poggi, Università di Bologna.

Under the reign of Empress Maria Theresa, the Duchy of Milan was a key center of Enlightenment discursive exchange and progressive social and economic reforms. It was home at the time to the influential journal *Il caffè* as well as to a number of famous learned women, including the mathematician Maria Gaetana Agnesi, whom Pope Benedict XIV had hoped to appoint to a mathematics chair at the University of Bologna. Apart from numerous civic reforms, the female sovereign enacted a sweeping modification of Milanese schools, establishing a science-based curriculum for secondary schools and the University of Pavia. It is conceivable, therefore, that the empress would seek to add Morandi and her anatomical oeuvre to the duchy's enlightened attractions. It should be noted, moreover, that Maria Theresa's sons, Peter Leopold and Joseph II, would both take a passionate interest in anatomical wax modeling, eventually amassing their own great collections.[33]

Other unconfirmed honors cited by Morandi's biographers include induction into Umbria's Academy of Art and the Society of Letters in Foligno. The Academy of the Gelati in Bologna did make her a regular member for her art.[34] Newly discovered epistolary correspondence discussed below also confirms that the University of Turin commissioned a series of wax anatomies from Morandi in 1756. Beyond Italy, the Royal Society of London solicited her to create waxes for its collection and invited her to visit. Catherine the Great also sought to bring her and her collection to Russia, but for a permanent stay.[35] Morandi undoubtedly received other commissions and invitations for which records no longer exist or remain to be discovered.

The Commission from Turin

A fascinating series of letters between Turin and Bologna from 1756 to 1758 sheds new light on the complex maneuverings that took place, at this time in Morandi's professional life, behind one patron's offer and Morandi's satisfaction of a large commission of wax anatomies for the University of Turin. Evident in this correspondence is Morandi's own hand in defending her interests and elevating the stature of her work in anatomy. The author of the letters was Ignazio Somis (1718–1793), Turinese court physician, professor from 1750 of medicine and medical theory at the University of Turin, and member of Turin's Academy of Science. The recipients were Eraclito Manfredi (d. 1759), professor of mathematics and hydrometry at the University of Bologna, who in youth was assistant to Giovanni Battista Morgagni in the Hospital of Santa Maria della Morte and a lifelong devotee of anatomical science, and Leopoldo Marco Antonio Caldani (1725–1813), anatomist and professor of practical medicine at the University of Bologna. At issue was the commission Anna Morandi received from the University of Turin to create a series of wax anatomical models of the sense organs for the Taurinense Museum.

On 10 November 1756, Somis wrote to his friend Manfredi in Bologna to ask that he make sure that the astronomer Petronio Matteucci sends the crate of wax anatomies for the "Royal University" to the mathematician and engineer Count Antonio Soliani Raschini of Brescello, who was a likely benefactor of the project. A year and a half later, on 29 March 1758, Somis corresponded again with Manfredi to say that he was writing to Signora Manzolini to inform her that all were agreed that she should not suffer a "grave inconvenience," and that she would in fact receive the 70 *gigliati* for her waxworks. What is plain from this letter is that in her own correspondence with Somis, Morandi had negotiated the price and protested the "grave inconvenience" to her of anything lower. A month later, on 12 April 1758, Somis changed inter-

Azzoguidi's copious praise of Morandi's skill in anatomy and consequent international celebrity and his deference to her on a question central to his own published study on the female reproductive system are unequivocal proof that her distinction did not derive solely from her exceptional position as a woman anatomist, or from her remarkable waxworks. Her right to claim throughout her scientific notebook the title of Anatomica, which I have interpreted in conjunction with her self-portrait as Lady Anatomist, originates in her proficiency in modern methods and theories of anatomy. That proficiency was nowhere more evident than in her expansive series on the sensory organs and the male reproductive system, the subjects of the next two chapters.

5

Esse est Percipi

Hands and Eyes

Sense and Sensibility

Innumerable tracts, books, and oral disputations by eighteenth-century Italian philosophers mapped with scientific exactitude the corporeal source of women's presumed intellectual inferiority and greater susceptibility to the senses. The most commonly identified capital of female deficiency was their cold, flaccid fibers by which merely sensual, as opposed to "sublime," knowledge feebly creeps along the nerves to the sluggish seat of the feminine mind/soul. Where men were seen to be able to govern their nervous sensations by acts of will, the least foreign stimulus and those innumerable resident feelings relentlessly fizzing inside women's bodies and minds as a rule overpower the feminine nervous system, which was deemed by many to have, quite literally, a mind of its own.[1] The eighteenth-century Venetian critic and writer Gasparo Gozzi summarized popular thinking on the subject in an article of his fashionable *Gazzetta veneta*: "Women's delicate, loose matter" heightens the acuity of their senses to such a degree that "the slightest breeze chills them" and "a minuet played on a flute or a violin instantly sets their knees to dancing."[2]

Typifying instead the (pseudo) scientific, empirical approach to the "woman question," the Paduan philosopher and cosmopolitan Antonio Conti also tunneled to the core substance of the female body—the fibers, where he pinpointed the material cause of women's perfect incapacity in the intellectual arts, the military, and government. He was, of course, seeking a conclusive, flesh-and-blood justification for keeping women out of those same dominant social arenas. In an epistolary essay that held sway for nearly two-thirds of the century, Conti argues that women's reproductive function leaves their fibers soaked and distended with milk and blood, causing the "blood [to rise] more slowly to women's brains . . . and depriving them of the tools that sustain the force of the mind and bring about . . . the formation of abstract ideas."[3] This lack of "vigor," especially of those most important fibers at the "oval center of the brain"—the seat, according to Conti, of the mind/spirit—naturally engenders women's erratic gestures and movements, and their extreme physical and psychological sensitivity.[4]

In his semiserious tract, "Croquis d'un dialogue sur les femmes," the Neapolitan author and diplomat to France Ferdinando Galiani takes eighteenth-century scientific misogyny to a new extreme. Reiterating the same essential list of physiological proofs—women's innately weak and deficient fibers as the cause of their acute physical and psychological frailty—he classifies the sex as inherently feeble, sick, and, indeed, a species of "deranged animals" always and utterly dependent on the pity and protection of men.[5]

Through her work and her very person, Anna Morandi invalidated these and other (male) authorities who continued to advocate ancient theories of women's inherent hyperbolic senses and sensibility. Her anatomical studies, based on a direct knowledge of the body's interior as opposed to the derivative notions of Gozzi, Conti, Galiani, and their like-minded brethren, implicitly contradicted the prolific eighteenth-century reauthorization of misogynist complexion and humoral theories according to the terms and tenets of experimental philosophy.[6] In her more than 240 pages of anatomical writings, Morandi identified no gender-based differences in the humidity and elasticity of the fibers, fascia, connective tissue, organs, nerves, or myriad other components of the bodies she dissected. Moreover, her series on the sense organs, the most famous and sought after of her oeuvre, illustrated their anatomical form and physiology in explicit relation to the body's experience of the external environment and signified a vigorous empirical defense of the senses' perceptive powers for cognition of the surrounding world.[7] Indeed, her detailed and evocative analysis of the anatomical structure and physiology of the sense organs exonerated them from the prevalent allegation that they are an unreliable conduit to knowledge. Perhaps even more critically, however, through her dissections, scientific writings, and wax sculptures of the sensory organs, Morandi physically and intellectually ruled the senses to which, as a woman, she ought to have been subject.[8] She herself, therefore, stood as a potent counterargument to misogynist leitmotifs of the burgeoning Science of Women. It was, in fact, by means of her own instruments of perception, her probing hands and conceptualizing eyes, that Morandi was able to elucidate the hidden structures and functions of the body.

Although Morandi did not explicitly join the vigorous sensationalist epistemological discourse of the Enlightenment, she sought to elucidate in text and image the anatomy and physiology of perception. Not surprisingly, she fixed specifically on the hand and the eye, those organs essential for dissection, sculpture, and a direct knowledge of the material foundations of human existence. Likened by Plato and Aristotle to none other than a block of wax on which are directly stamped our impressions of objects, sensory perception became during the era of the "new" science and philosophy a contested medium for knowledge of the external world. At the heart of the philosophical and scientific inquiries of such "moderns" as Descartes, Locke, Newton, Berkeley, Hume, and Kant was the question of the function and the limitations of sensibility for cognition.[9] While Morandi was expressly concerned with sensory vision and kinesthetic perception, her interpretations of the body's experience of the external world had indisputable implications for broader epistemological questions as well as the growing eighteenth-century science of the mind.

De Visu: *The Eye Unbound*

The eye by long use comes to see even in the darkest cavern: and there is no subject so obscure but we may discern some glimpse of truth by long poring on it.

GEORGE BERKELEY, no. 368 in "A Chain of Philosophical Reflections"

On her first appearance as celebrity disputant at the Carnival Dissection in 1734, Laura Bassi faced her academic rival, the professor-anatomist Domenico Gusmano Galeazzi, across the corpse of the dissection subject and the Archiginnasio Anatomy Theater to contest a prevailing question among contemporary Bolognese and European philosophers: the nature of sight. Manifests wrapped around the courtyard pillars of the Archiginnasio had primed the teeming audience for the debate, "De visu."[10] It was thus with certain expectancy that university doctors, religious leaders, and prominent foreigners took their seats to witness the celebrated Bassi return to and, it was hoped, expand the arguments she had triumphantly upheld two years earlier in defending her forty-nine philosophical theses, the final rite of passage before she became the first woman awarded a degree by the University of Bologna.

Bassi had established herself clearly as a young Newtonian opticist at that ostentatious ceremony held in the seat of the Bolognese government, the Public Palace, on 17 April 1732.[11] Her disputations under the rubric "De anima" (on the mind/soul), in particular, focused on contemporary theories in optics. She reiterated key findings of the famous Newtonian experiments on the light refraction and the immutability of colors carried out by the young Francesco Algarotti and his influential teacher Francesco Maria Zanotti at the Institute of Sciences in 1728.[12] Indeed, the philosophy of seeing and the physics of sight gripped the scientific imagination of the institute's professors at the exact time of Bassi's rise. Only a few months before her performance at the Public Anatomy in 1734, Zanotti had presented at the institute new experiments on the refrangibility of light rays, pitting a common spyglass against a proper Newtonian scope.[13] Bassi unmistakably staked her claim to intellectual eminence on this new branch of knowledge. Algarotti would reinforce her exceptional standing as an original newtoniana a few years later, in 1737, with the publication of his bestselling *Newtonianismo per le dame* (*Newtonianism for Ladies*). Bassi served as a model and a symbol of the dissemination of optical science, which Algarotti sought to broadly expand with his popular exposition of Newtonian optics.

Although the specific arguments put forth by Bassi in her disputation with Galeazzi at the 1734 Carnival Dissection are not known, it is fair to conjecture based on eyewitness accounts that the woman philosopher spoke about the anatomy of the eye in ways well known to eighteenth-century scientists, in which select parts facilitate the mechanics of sight.[14] How, for example, do we apprehend a single object although we look with two eyes? She would doubtless have reiterated and explained the analogy frequently drawn at the time between the operations of the eye and a camera obscura. She would likely have discussed how objects transmitted to the eye upside down are seen by the mind right side up; how the crystalline humour, or lens, serves to focus light rays on the retina, adjusting for the distance of the object observed; how the muscles that surround the eye move it up, down, and in all directions. But the crux and substance of her argument would have been a learned encomium to the English physicist and his theories of light, specifically how what we see is actually the result of a "heterogeneous Mixture of Rays some of which are constantly more refrangible"—so central to local intellectual discourse at the time and to her own extraordinary standing in the public eye.[15]

As a Bolognese citizen with a pulse, Anna Morandi certainly knew of these events, both actual and textual. However, as a striving "professor" of modern human anatomy, she undoubtedly aimed to understand and to integrate in her own studies the relevant anatomical fine points. As we have seen, Morandi and her husband shared a driving ambition to establish themselves as leading anatomists within the eminent local tradition of de' Liuzzi, Vesalius, Malpighi, Morgagni, and especially Valsalva. Their specialization in the organs of sense reflected a desire not only to supersede the work and local acclaim of Antonio Maria Valsalva on the anatomy of the ear but to contribute decisively to the prominent contemporary discourse on the philosophy and mechanics of perception for which Laura Bassi, Francesco Algarotti, and other "modern" *accademici* had won local and international acclaim. It is no surprise, then, that Morandi began her anatomical notebook by focusing on the eye, the organ of sight. Its first twenty pages inventory and explain her prolific sculptures of intact and systematically dis-integrated eyes.

More than a map of the components and composition of the organ, her extensive series on the eye, as throughout her oeuvre, also focused on the active function or experience of the eye. The numerous "looking eyes," shown either naturally embedded in the face or *ex situ*, as well as the dissected parts visualized on myriad tables and in intricate detail, were animate. Morandi set atop tiny wooden pedestals seeing eyes that gazed from partial, though vital, human faces (fig. 51). These served to depict the living eye functioning naturally within its orbit. Arranged on numerous octagonal tables like numbers on a clock face, whole and extricated eyes also peered decisively in different directions, delimiting the eye's range of motion (fig. 52). Other octagonal slabs displayed dissected lids, corneas, retinas, tear ducts, glands, and the muscles and nerves that surround the eyes (figs. 53–56).

Her visualizations of the eye's more tenuous parts also included microscopic components, including ones she was the first to discover. No matter how parsed or minute the pieces of the eye on display, all were dynamic and served to theorize, albeit rudimentarily, how the eye operates and sees. There was special attention, especially in the beginning pages of her notebook and models, to the numerous eye muscles, their points of attachment and their influence on eye movement. Morandi likewise concentrated on the mechanical cooperation between the parts of the eye that facilitated sight. When describing the lachrymal gland, for example, she highlights the practical function it served in lubricating and thus moving the eye and the lid:

> At the ciliary puncta, located at the top of the grooves at the internal part of the tarsi, there are openings or excretors of the sebaceous gland of the bilateral form that serve the lymph that facilitates the movement of the lid and maintains the moistness of the eyeball.[16]

As she progressed in the series from indexing the muscular structure to a more complicated analysis of the machinery of the eyeball, her descriptions and wax facsimiles also become more assiduous and intricate:

> The second layer common to the eye is called the choroid. This is divided into two portions, the one major the other minor. The major, called the choroid, is

FIGURE 51. Anna Morandi, partial face and eye, wax and bone. Courtesy of Museo di Palazzo Poggi, Università di Bologna.

FIGURE 52. Anna Morandi, eyes, wax. Courtesy of Musei di Palazzo Poggi, Università di Bologna.

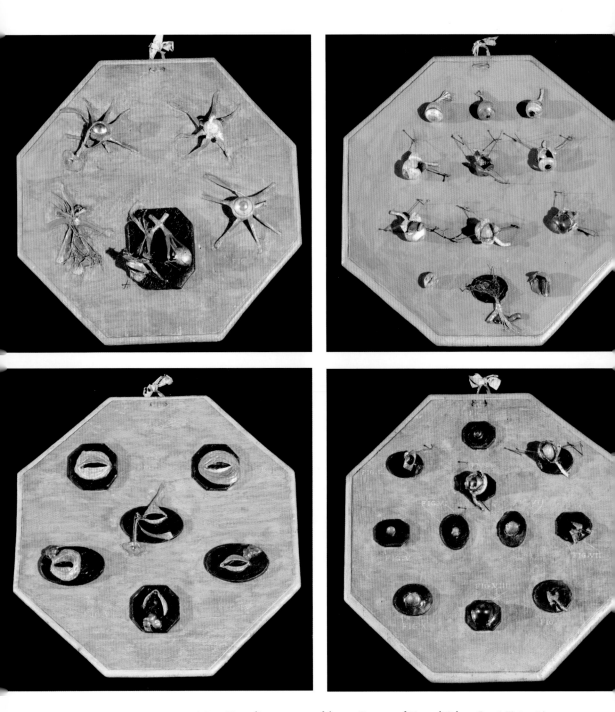

FIGURE 53–56. Anna Morandi, components of the eye. Courtesy of Museo di Palazzo Poggi, Università di Bologna.

united to the internal surface of the opaque cornea and, notably, is attached at
the union of the two tunicas, opaque and transparent, by way of many fibers
that form a ring called the ciliary circle. The minor portion of the uvea at the
anterior makes up the colored part of the eye called the iris, in the middle of
which is a round opening called the pupil. This minor portion of the uvea is
formed in its internal part by many fibers, some circular, some radial, which
together form the ciliary process.[17]

At discrete moments we are, in fact, able to enter Morandi's laboratory and
perceive the dissected eye as she saw and felt it. In her description of the sixth
figure shown on the seventh display table, she explains that it served to demonstrate "how the crystalline lens is naturally so delicate that if you touch
it and lightly press it with the tip of a finger, it immediately collapses and
distorts."[18] Behind the lens she points to the vitreous humor, whose substance,
she maintains, could best be seen "by looking through it in the light."[19]

Keen to exhibit her anatomical expertise and the precision of her method
and resultant observations, at several points Morandi overtly challenges other
anatomists' claims about the eye's anatomy. She states, for example, that "contrary to the opinion of some authors, it is important to note that the transparent cornea is never joined to the opaque, as I have shown in all the tests I have
been able to conduct."[20] Among her most momentous and oft-touted contributions to the study of human anatomy is her discovery of the course and
attachment of an oblique eye muscle. She begins her description of this muscle
by acknowledging what was commonly known to anatomists, only to point
out the insufficiency of the account and her own more complete analysis: "the
oblique inferior muscle not only attaches to the nasal apophysis of the maxil-
lary bone, as the authorities agree, but with the bone opened, one sees that the
muscle proceeds and attaches itself to the lachrymal sac. This was discovered
by me in my observations and I have found it always to be constant."[21]

She also created a colossal eye (now lost) six times the actual size, depicted on
two tables. Although not listed as part of the series, it was registered with Count
Girolamo Ranuzzi's purchase of her complete collection in 1769. As with the
colossal ear she also made, the enormous eye permitted close study of the minute
parts and contributed its own new perspective on the organ of sight.

Plainly implicated in Morandi's representation of the mechanics of sight
was the scientist-artist's own penetrating vision. Both specular and mnemonic,
the realistic wax eyeballs not only replicate and return the viewer's gaze, they
implicitly recall the original visionary, Morandi, who held and beheld these
eyes. The ordered breakdown of the body, in this case the eye, into ever more
particularized figures summons to mind the delicate cooperation between the
scalpel interrogating the body's incalculable layers and parts and the anato-

Morandi incorporates with her graphic facsimile a recurrent theme of Enlightenment epistemology, that pleasure and pain are paradigmatic of sensation.[26] Here, as elsewhere in her oeuvre, she represents the body as experience (a term that, we recall, signified during the Settecento both lived reality and scientific experiment) and, I would add, visually manifests both the cognizance and the memory rooted in the senses. Through artistic imagination and scientific technique, she resuscitates the limp, colorless, dissected hands of a cadaver, transforming them into realistic corporeal manifestations of pleasure and pain. Her own "corporeal imagination"[27] allowed her to re-member in wax intact, reactive fingers that act as synecdoches for the whole, living and feeling body. She created a compound metaphor for subjective corporeality and empirical proof. Indeed, it is not unreasonable to conjecture that Anna Morandi modeled the original "feeling hands" in the series on her own nimble and knowing hands experienced in wielding scalpel and forceps, delving into the bodies of the dead, and shaping and sculpting resurrected bodies in wax.

The scientist was quick to restrain these original, evocative signs of touch through her analytic method of representation and interpretation. With words and image she pinpoints the anatomical structures enabling the sensory perception of the hands. In her notes, she explains that nature concentrated nerve endings in the hand in order to signal the appropriate response to the external environment. A series of five meticulous wax figures succeeded the two "feeling hands," peeling away, layer by layer, their intricate structure until they touched bone (fig. 58).

In the crowded Third Table of the series, in which she demonstrates the hand stripped of its coverings to show the complex infrastructure of muscles, tendons, fascia, and, most important, nerves, she states:

> All of the nerves that have been identified extend to the summit and apex of the fingers and here multiply into very copious and minute branches, more than in any other part of the body. And because of nature's prolific multiplication of the nerves, sensation is rendered more acute and delicate here than anywhere else.[28]

Her exhaustive examination of the parts of the appendage that enable sensation concludes by describing the extent to which the neurovascular structures that facilitate feeling reach into the diaphanous layers of the hand's skin:

> [Here one sees] the cutis (derma or true skin) with its papillae, or cutaneous glands, which are those prominences surrounded by small furrows that form the interstices between the one and the other. All of the blood vessels and nerves end in this membrane of the cutis, while the cuticle (epidermis) is wholly deprived of the same and therefore without any sensation, and thus serves no other function than as a protective cover for the cutis itself.[29]

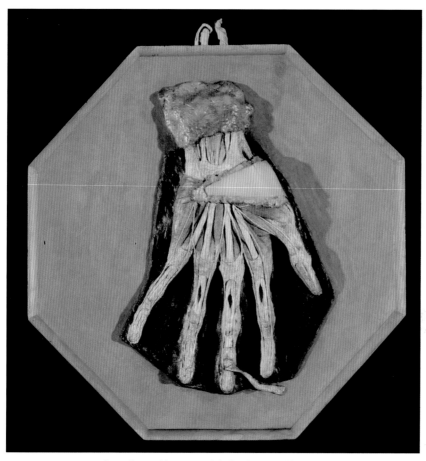

FIGURE 58. Anna Morandi, tendons and ligaments of the hand, wax and bone. Courtesy of Museo di Palazzo Poggi, Università di Bologna.

Morandi's wax figures and companion notes on the anatomy of the eye and the hand illustrate Francesco Algarotti's theories on the sense of touch and its relation to sight, those two essential cognitive faculties that, he asserts in the third dialogue of his *Newtonianismo per le dame*, "together shake hands and accompany each other in the formation of our ideas."[30] While Morandi implicitly privileges the sense of touch in her distinctive rhetorical and visual treatment of the hand, Algarotti is overt in assigning superior perceptive powers to the hand. The eyes, he argues in a lengthy dialogue between his mouthpiece, the Chevalier, and the novice Marchioness, are incapable of interpreting accurately such essential attributes as distance, number, and extension, and myriad qualities like softness, hardness, sharpness, smoothness, and so on, without the aid of touch. The eye cannot assess physical reality, in Algarotti's view, unless it is quite literally given a hand:

This sense of touch, which is much stronger than sight, has constantly advised us in the ordinary manner of seeing, that an object is one alone and we, by means of long habit, join the idea of a single object to the two sensations of it. . . . The ideas derived from sight are, in contrast to [touch], like four strokes of a pen compared to a fine relief. We have the example of a Sculptor who, while blind, nonetheless sculpted by touch some portraits that were quite acceptable. . . . Touch provides the fantasy much clearer and more precise ideas than sight can give. . . . On the other hand, what would we understand and do without touch. Incapable of judging the size, distance and form of objects (as Berkeley had declared, who more than any other has considered the metaphysics of vision . . .), incapable, I say, of using our eyes, we would be invited by sight to knowledge and pleasures that we could never attain.[31]

The hand that touches thus adds sensible meaning to what we see. Devoid of touch, we would be incapable of interpreting our experience, and indeed, for Algarotti, we are incapable of truly seeing, since everything seen would remain unqualified and at the level of abstraction. Algarotti also stresses the lasting beneficial influence of touch on our interpretative faculties. Touch imprints sensual memory on the mind, bolstering subsequent sensory assessment and the right physical response to the external environment. The reiteration of touch and the accumulation of hands-on knowledge serve to reinforce our interpretive powers. Anna Morandi's series on the hand visually and scientifically echoes Algarotti. Her whole, "feeling" left and right hands are shown responding appropriately, even passionately, to opposite, outside physical stimuli. As we have seen, she then peels back the skin of the fingers and the palm of the hand to show the intricate nerve endings, the subtle blood supply, the tendinous pulleys and bony levers, indeed the skin itself, that endow the hand with its distinct and practical ability to apprehend what the eye cannot.

Morandi's wax effigies of the hand and highly suggestive writings on the sense of touch in her notebook must also be seen as a means of establishing her elite lineage as an anatomist. Her focus on the hand confirms her descent from the science's two most venerated masters, Galen and Andreas Vesalius. Both privileged the hand, in accordance with Aristotelian doctrine, as the "instrument of instruments," the *organum organorum*. Galen launches his *On the Usefulness of Parts of the Body* with a prolonged exposition of this the most noble of the body's parts, perfectly designed by Nature to raise humankind above all earthly creation. The complex and exceptionally dexterous append- age with its opposable first digit permits humans alone to make tools both for survival and for science and the creation of art. As Nancy Siraisi has noted, Galen dedicates the entire first book and most of the second of his anatomical teleology to the hand. He dwells particularly on the innumerable fine muscles and tendons that permit the fingers' difficult movement.[32]

Although Vesalius does not give isolated treatment to the hand in the *Fabrica* but instead discusses its anatomical structure in relevant sections of four of the seven books, in other more provocative ways the author reveals himself to be a "cultist of the hand," as William Heckscher has described him.[33] Vesalius intentionally distinguishes himself from past professors of anatomy by practicing anatomy with his own two hands. His real and iconographic break with convention centers on this literal hands-on knowledge of human and comparative anatomy. The title page to the *Fabrica*, in which he is represented reaching deep into the abdomen of a female cadaver, is an affront to traditional anatomy and its hierarchic profession, emblematized on the title pages of prior atlases by an enthroned lecturer, a ground-level demonstrator, and a lowly barber-surgeon, who alone touches the cadaver.

Reaffirming his image as the modern *chirurgus*, who, as the name denotes, performs his art specifically by manual operation, is the 1542 woodcut portrait of Vesalius showing him grasping with his own two hands the exposed elbow joint and the resected flexor digtiorum superficialis of the forearm and hand of a curiously upright cadaver (fig. 59). The famously diminutive anatomist, standing in profile alongside the outsized, flayed arm of the suspended corpse, appears to be a kind of David to an anatomical Goliath, underscoring the audacity of Vesalius's challenge to professors of anatomy: know thyself, *gnothi seauton*, not through authorized texts but through the direct touch and experience of the body.

Is Anna Morandi aiming, as the new anatomist of the hand, to be the Vesalius *redivivus* of mid-eighteenth-century Bologna? Numerous others before her asserted that claim via portraits of themselves dissecting the forearm and hand, from Giulio Casserio to Volcher Coiter and the Dutch anatomist Nicolaas Tulp, who was immortalized by Rembrandt.[34] Certainly she seeks to buttress her place within the modern Vesalian anatomical tradition through her focus on the hand. The singular attention to the perceptive powers of the hand that she represents in wax and her implicit engagement with contemporary philosophical discourse in her writings on sense perception would, moreover, seem to extend the reach of her practice, of her own anatomizing hand, to the academic cathedra of Bologna and Enlightenment Europe.

ANDREÆ VESALII.

FIGURE 59. Woodcut portrait of Andreas Vesalius for frontispiece, *De humani corporis fabrica* (1542).

6

Beneath the Fig Leaf

The Male Reproductive System and Genitalia

Upon his return to Venice from a tour of the city of Bologna in the spring of 1817, Lord Byron dispatched two letters in two days describing the wax models of the male and female sex organs sculpted by Anna Morandi, on display in the Department of Human Anatomy at the University of Bologna.[1] On 3 June he wrote to his half-sister and confidante, Augusta Leigh, the more restrained account of the "female professor of anatomy, who has left there many models of the art in waxwork, some of them not the most decent." In his letter the following day to his publisher, John Murray, Byron was far less allusive:

> I forgot to tell you that at Bologna—(which is celebrated for producing Popes—Painters—and Sausages) I saw an Anatomical gallery—where there is a deal of waxwork—in which the parts of shame of both sexes are exhibited to the life—all made and molded by a *female* Professor whose picture and merits are preserved and described to you—I thought the male part of her performance not very favorable to her imagination—or at least to the Italian Originals—being considerably under our Northern notions of things—and standard of dimensions in such matters—more particularly as the feminine display was a little in the other extreme—which however is envy also as far at least as my own experience and observation goes on this side of the Alps—and both sides of the Apennines.[2]

Byron's letters express the fascination with which European travelers to Italy continued to regard the anatomical waxes sculpted by the Bolognese woman anatomist and modeler, even after her death in 1774. From the mid-eighteenth century to the early nineteenth, numerous eminent tourists in that country, including the French astronomer Joseph Jérôme de Lalande, the American

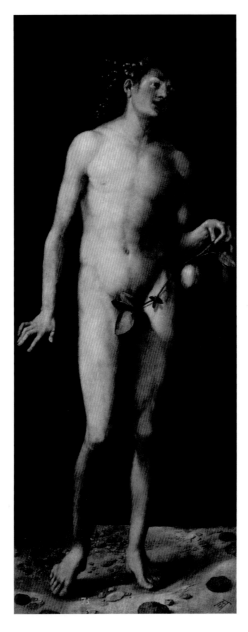

FIGURE 60. Albrecht Dürer, *Adam* (1504). Oil on panel. Museo Nacional del Prado, Madrid.

professor of medicine John Morgan, and the British woman author Hester Lynch Piozzi, recorded their visits and reactions to Morandi's well-known waxworks, a featured stop on the Bolognese tour.[3]

However, Byron's suggestive missives also tell of the prurient appeal of these wax anatomies for early modern viewers (and no doubt later viewers as well). His emphasis on the *female* sex of the anatomy professor further reveals this to be what most excited that appeal. Expertly sculpted with the same feminine hands and eyes that had sectioned the parts of cadavers on which they were based, these anatomical figures graphically embodied Anna Morandi's intimate and exceptional knowledge of those most secreted realms of what Andreas Vesalius called the "human corporeal fabric," from tendons to lymph nodes, the brain to the extreme reach of plaited arteries and veins. Morandi's proficiency in the generative parts was thus all the more provocative having been expertly unveiled by a woman.

At the height of her career as anatomist and anatomical modeler, Morandi sculpted a comprehensive series demonstrating the male urogenital system and genitalia, comprising a stunning 22 wax models and 47 pages of companion scientific notes. Historians have virtually overlooked this focal point of her oeuvre, no doubt in part because of extensive gaps in the primary documentation. In contrast to the rest of her collection of wax models, which is largely intact, nearly all the figures in this series are lost.[4] However, enduring disregard for Morandi's radical recasting of the gender roles standard to the anatomical dissection scene certainly has also been influenced by a persistently narrow view of women's position in the realm of the body during the early modern age.

FIGURE 61. Ercole Lelli, female nude, *Eve,* wax. Courtesy of Museo di Palazzo Poggi, Università di Bologna.

Woman's Place in the Dissection Scene

The gaze of the female subject in the traditional anatomical scene was of three principal types: the humiliated, taciturn look of the anatomical Eve; the fixed, dead stare of the female cadaver; and the voluptuous, aroused trance of the anatomical Venus, produced in abundance in Florence at the end of the eighteenth century. The archetypal Eve sculpted by Morandi's forerunner in anatomical ceroplasty, Ercole Lelli, is shown, as we have seen, in classic fashion with bowed head, downcast eyes, and a defective veil of flowing hair that cannot hide her shameful nakedness (fig. 61). She stands together with Adam at the head of a series of eight life-size anatomical figures progressively demonstrating the muscular and skeletal systems within an allegorical tableau of sin and mortality. The intact female figure, whose skeletal counterpart emblematically completes the series by brandishing an iron sickle (fig. 62), connotes both fallen femininity and moral recompense against the criminal body. The title page to Vesalius's *Fabrica* manifests even more starkly the debased position of woman standard to the early modern dissection scene, that of inert speci-

FIGURE 62. Ercole Lelli, female skeleton with sickle. Courtesy of Museo di Palazzo Poggi, Università di Bologna.

FIGURE 63. Andreas Vesalius, title page of *De humani corporis fabrica* (fig. 46); detail. Courtesy of Becker Medical Library, Washington University School of Medicine.

men beneath the scientist's deliberate touch and gaze (fig. 63). As Katharine Park has incisively observed, "Vesalius embodied his own vaunted reform of anatomy in a scene that shows him exposing the entrails of a female cadaver to an unruly band of male colleagues and students."[5]

Yet the marked ambition of the anatomist to take possession of, lay bare, and thereby discipline the *secretum mulieribus* exhibited in this canonical image reached a hyperbolic intensity in the wax anatomical Venus. These popular figures, complete with romantic hair, glass eyes, and jewelry, offered a vision of superficial feminine sexuality as well as access to the hidden depths of the female sex. Spectators were invited to behold the successive removal of layer on corporeal layer in the mode of an anatomical undressing: off came the breast-plate, the superficial muscles, the deep muscles, and the lower abdominal organs (fig. 64). The climax arrived at the disclosure of the naked core, the gravid uterus with its tiny fetus in view. Far more rewarding than scopophilia, each separate anatomical component could be physically extracted and possessed.

Feminist theorists and historiographers Ludmilla Jordanova, Elaine Showalter, and Karen Newman have performed notable *analytical* dissections of the symbolic position and meaning of the early modern female anatomical subject, in particular that most verisimilar variety wrought in colored wax.[6] Jordanova and Showalter have explicated the cultural and political conno-tations of the Florentine Venuses, while Newman has sought to show the link between current anti-abortion obstetrical representation and the wax uterine models of the eighteenth-century Bolognese School and Museum of Obstetrics.[7] In each case, the author traces the poses, the naturalistic shades,

FIGURE 64. Clemente Susini, *Venus,* wax. Courtesy of Museo di Palazzo Poggi, Università di Bologna.

and the contours of the wax models back to the commanding gaze of the scientist and that of his presumably male audience. This expectant male onlooker was, according to Jordanova, "intended to respond to the model as to a female body that delighted the sight and invited sexual thoughts."[8]

Building on Jordanova's critique, Showalter contrasts the fetishistic interest in the anatomized female body evidenced by the production of the wax Venuses with the absence of a complementary interest in male anatomy. Overlooking such widely circulated texts as Regnier de Graaf's 1688 *Tractatus de virorum organis generationi inservientibus* (Treatise concerning the Generative Organs of Men),[9] as well as the numerous male wax models displayed alongside their female counterparts in Bologna and Florence, Showalter avers that "there were few overt cultural fantasies about the insides of men's bodies, and opening up the man was not a popular image." She attributes this disparity to men's control over anatomical science as well as their *gynophobia*: "[Men] open up a woman as a substitute for self-knowledge, both maintaining the illusion of their own invulnerability and destroying the terrifying female reminder of their impotence and uncertainty."[10]

Newman also disregards relevant representations of the male body, in particular the reproductive male body, in her study *Fetal Positions: Individualism,*

Science, Visuality, which focuses on eighteenth-century figures of the gravid uterus and female reproductive system created for the first Bolognese Obstetrical Museum. Newman interprets these figures as a "series of disembodied wombs," separated and decontextualized from the whole, living female body, and thus as antithetical by design to the "always male fetus . . . conceived of as preformed, a fully fashioned though tiny adult that simply grew in size."[11] Her presentist objectives lead her to construe the Bolognese obstetrical models as prototypes of contemporary images of the reproductive female body in anti-abortion propaganda and, as such, to "constitute crucial political knowledge for the present."[12]

Newman's argument falters on thin historiography based on an incomplete and skewed consideration of the facts. She offers a cursory account of the eighteenth-century Bolognese cultural context and the individual and institutional patronage that gave rise to the wax obstetrical models. She makes use of spare secondary sources on the history of the obstetrical models (studies whose validity she also tends to discount if they undermine her claims),[13] while she overlooks crucial primary documents that illumine the Bolognese Senate's and Pope Benedict XIV's benefaction of the obstetrical models for the public health and welfare. No reference is made to Professor Giovanni Antonio Galli's stated objectives in commissioning the models, including his directions regarding the subject matter and mode of construction, and his actual use of them in his practical courses for midwives and surgeons, which I discussed at length in chapter 3.[14] Finally, she grossly misinterprets "as not-so-veiled sadomasochism" the presentation mode of the Bolognese wax anatomies, sculptures she also incorrectly slots into a single aesthetic and ideological enterprise with the Florentine and Viennese models that were made later, by different artists and scholars and physicians, and for distinct aims. This conjectural retrospection for the roots of current imaging of the reproductive female body, a use of the past fraught with risk even when founded on painstaking documentation, overlooks the extensive written description and scientific analysis of the Bolognese wax models by the anatomists/artists who made them, including, of course, Anna Morandi. As this written record makes manifest, far from being wrought from aggressive or destructive impulse, as Newman suggests, the scrupulous wax anatomies of the structure and function of vital organs, appendages, and systems (muscular, reproductive, nervous, circulatory, osseous, etc.) of the living body were practical teaching tools as well as epitomes of the body's anatomical design.[15]

Each of the studies discussed above posits an unqualified dichotomy of the fully realized male subject (anatomist, spectator, and indeed fetus), his presumed agency and desire, and a conversely submissive and ontologically deficient female corporeal object. Yet, in her discerning critique of predominant feminist interpretations of the symbolic and cultural significance of the wax Venuses, Mary Sheriff indicates the historical and theoretical insufficiency of only considering "men as they look at women's bodies." Sheriff points out that "women appear . . . only insofar as they are objects for the male gaze and are never considered looking themselves."[16] To Sheriff's analysis, I would add that, ironically, the three canonical interpretations of the wax Venuses cited here represent a disembodied historical and cultural critique. We are not told on the basis of documents essential information about who made the waxes, why and for whom they were made, or who actually viewed them. Moreover, absent is any acknowledgment of the authority and subversive power of the female anatomist, scientist, artist, and spectator during the period who looked on and at times possessed expert knowledge of the body, including the sexed body.

"As who dare gaze the sun"[17]

In the new anatomical scene that Anna Morandi constructed and occupied, woman was no longer the docile anatomical object of discovery, the vacant-eyed cadaver at the center of the thronging theater, splayed and rent by the master anatomist. Rather, as Morandi's wax self-portrait makes manifest, she herself is the master anatomist, who directs her own knowing gaze on and inside the parts of the body she uncovers, including the brain, the sense organs, and indeed, the male sex.

Gianna Pomata's recent analysis of the reconceptualization of the female anatomy and generative functions from the second half of the seventeenth century and of the concurrent rise in interest in the question of sexual difference serves as a theoretical backdrop for my interpretation of Morandi's position and contributions in the sphere of reproductive anatomy. Pomata provides a crucial, document-based account of the burgeoning scientific currents that sought to supplant Aristotelian theories and their scholastic application with an experiential method and reassessment of the parts and processes of human generation.[18] This historical revision of the scientific revolution bolsters the well-regarded criticism by Katharine Park and Robert Nye of Thomas Laqueur's influential thesis of a radical shift in conceptualizations of sexual difference that occurred during the European Enlightenment. Laqueur claims that this was the first age to jettison the monolithic "one-sex *male* model," in relation to which woman was seen as derivable and deficient, and to put in its place the prevailing "two-sex model," in which the sexes are conceived as fundamentally distinct.[19] Focusing on two interconnected scientific innovations, Pomata disputes Laqueur and especially feminist historiographers who argue with him that "the Scientific Revolution did not displace the dogma of male superiority nor shed light on the problem of sexual difference."[20] The widely accepted redefinition of "female testicles" as ovaries and the discovery of spermatozoa by Antoni van Leeuwenhoek during the mid-seventeenth century are clear evidence, she argues, of a robust "anti-Aristotelian view of sexual difference"[21] well before the Enlightenment. Ovist and spermist theories of the seventeenth century, despite their essential opposition, manifest a shift away from received methods and ideas regarding human generation as well as a clear design to uncover the discrete and indispensable reproductive functions of each sex.[22] Moreover, according to Pomata, as a result of the crucial and distinct generative role attributed to ovaries, early modern medical writings frequently express the view that "both sexes are necessary to reproduction and that each is perfect in relation to its own function."[23]

Through her systematic and detailed analysis of the parts and functions of the reproductive male body, Anna Morandi contributed to the discourse of sexual difference as it evolved during the eighteenth century. Indeed, Morandi unmistakably allied herself with those progressive early practitioners of the empirical method in the realm of the body to whom Pomata calls attention. With Marcello Malpighi, Regnier de Graaf, Antonio Maria Valsalva, Giovanni Battista Morgagni, and other anatomical empiricists whose work she studied, Morandi looked anew at the body and explored its deepest depths and microscopic elements literally firsthand in order to know and accurately convey its design. However, notwithstanding the unmistakable merit of her analysis, Pomata's generalized critique of the feminist historiography of the age and her conclusions that "the assumption of male superiority . . . had become . . . an open problem" in seventeenth-century medical discourse to such a degree that this period gave rise to a "new positive perception of the female body"[24] elide the vast early modern cultural discourse on the subject of female inferiority, particularly that especially relevant body of literature Luciano Guerci has so aptly designated "scientific misogyny."[25] As I aim to show, Morandi stood at the intersection of the two dominant scientific and discursive currents in contention in Pomata's analysis: the progressive empirical mode of scientific analysis of the body and sexual difference, and the age-old "dogma of male superiority"—seemingly contrary dispositions that in fact often coalesced in the same "masterwork."[26] Morandi's method and practice of anatomical science derives from and furthers the scientific re-vision of the body that Pomata elucidates. However, her controversial position as woman anatomist, for which she came under explicit and implicit attack, as well as the tactical partiality of her visual and written account of human reproduction also serve to shed light on the robust "scientific misogyny" that persisted and indeed thrived through the course of the Enlightenment.

Morandi delved deep beneath the fig leaf to uncover the intricate structures and corresponding functions of "the male generative parts." She did so in direct defiance of contemporary rules governing the art of anatomical design. Influential arbiters of Bologna's cultural standards publicly disparaged, in letters, tracts, and staged comedies, *artisans* who asserted an equal claim to scalpel and chisel, the ideal exterior and the unruly interior of the body. More to the point, numerous leading voices within the cultural establishment, as has been seen, explicitly denounced the training of artists in anatomical science beyond what was necessary for "being able to perfectly and completely represent the form of the male nude."[38] Morandi thus transgressed the established borders between art and science no more egregiously than with her meticulous, verisimilar sculptures of the dissected male sex.

Yet, while Anna Morandi engaged this provocative subject to vie for a place in the society of illustrious anatomists, she allied herself with authors who, unlike Petronio Zecchini and like-minded apologists for the "science" of female deficiency, sought to reconceive the human corporeal fabric and sexual difference through impartial hands and eyes.[39] In contrast to the anatomist in the dissection scene revised and modernized by Andreas Vesalius in his portrait on the frontispiece of the *Fabrica*, Morandi did not establish her authority through an act of subjugation and conceptual and rhetorical violence. As Park has deftly described: "by choosing a female cadaver for his frontispiece; by placing that single female body in the middle of an unusually large and rowdy male crowd; and by placing it, and himself, in a position calculated . . . to magnify the sexual element," Vesalius, and his many early modern devotees, ascribed to the male superior formative virtues for the generation of offspring, while proffering "a corresponding denigration of the women's part."[40] The "Maker of Things," according to Vesalius, "so constructed humans in the beginning that one [man] would put forth the supreme reason for the beginning of the fetus while the other [woman] would receive it and nurture and foster the fetus."[41]

Morandi's study of the reproductive male body avoided comparative analysis and maintained an implicit indifference to sexual difference. As will be shown, while self-proclaimed "modern" practitioners of anatomy persistently acted as "blindfolded gladiators"[42] brandishing received theories to defend conventional misogynist views, Morandi, in fact, practiced a more purely empirical study of the male anatomy based, as Vesalius himself required but did not always practice, on the accurate description "of each item as it appears in dissection."[43]

With such practitioners of the "new" anatomy as Regnier de Graaf, Marcello Malpighi, and Giovanni Battista Morgagni, whose work she knew and in some cases cited and even corrected, Morandi disclosed the particular parts, structures, and processes of the reproductive male body.[44] These authors, as Gianna Pomata has shown, advanced the discourse on sexual difference by ascribing greater importance to women's distinct reproductive system and functions.[45] The Dutch anatomist de Graaf (1641–1673) substantiated William Harvey's claims of 1651 that "ex ovo omnia" (all living things come from eggs) by elaborating the function of ovaries,[46] identifying their follicles (which he erroneously believed were eggs), and mapping the passage of eggs through the fallopian tubes to the uterus. From his microscopic studies of chick embryos, Malpighi (1628–1694) identified, among other parts, the embryonic blastoderm, developing blood vessels, somites, and neural grooves. He also amended the theories of de Graaf, with whom he had an extensive correspondence, by distinguishing follicles from eggs.[47] Morgagni (1682–1771) conducted extensive anatomical demonstrations to arrive at his conclusions with de Graaf that female ovaries produce eggs from which, through fertilization by semen (which he and de Graaf incorrectly believed took place in the ovaries), offspring are produced.[48]

Although Morandi makes no explicit reference to the eminent Paduan naturalist Antonio Vallisneri (1661–1730), it is probable that she also had some familiarity with his ideas on human generation, which were at the time widely discussed among Bolognese academicians. During his lifetime, Vallisneri was the Chair of Practical and Theoretical Medicine at the University of Padua, physician to the Hapsburg Emperor Charles VI, a prolific correspondent with academicians throughout Europe, and author of numerous tracts and books on subjects ranging from earth and planetary sciences to zoology, hydraulics, and anatomy. In 1721, he published a vast *Istoria della generazione dell'uomo, e degli animali* (History of the Generation of Man and Animals) that changed how anatomy was seen and practiced. By means of systematic observation, experiment, and reason, he aimed to determine the validity of "two principal opinions of the century": "whether the Glory of generation was given to certain worms that frolic in male semen . . . or to the egg of the female in which the whole of the machine of the animal is enclosed awaiting only the motion, the vital push or the spirit of the male fecundator."[49] Although critical of the competitive debate on human generation in which there are those "who give all the glory of generation to the Woman and those who give it all to the Man,"[50] he came down firmly on the side of the egg. Vallisneri admits to having seen the seminal worms that "the very diligent Leeuwenhoek with his most rare and marvelous microscopes was among the first to discover,"[51] but he avers that the more important question is, for what purpose did the Supreme Author make them?[52] He concludes that semen and its component parts serve to facilitate the generation of offspring, "by introducing the principles of spirit, motion and life."[53] The perpetually wriggling spermatic worms specifically acted to keep the viscous semen from becoming stagnant and excessively dense while stored before ejaculation.[54]

Undertaking an expansive review of the uniform laws of generation in nature by comparing the birth of humans to plants, insects, birds, and animals, Vallsineri reiterates in hundreds of pages of scientific similes the same basic generative scenario: in near-exact likeness to an ant (*formica*), for example,

> by the worm of the Man the Egg of the Woman becomes fertilized. The egg dilates and readies itself to leave the ovary. The ovary emits the egg of the woman, which falls into the uterous as into a nest perfectly proportioned within its burrow. There, incited by the heat of the uterous and quenched by a most benign fluid (lymph) it grows and is made ready to receive extra food from the Mother. The Mother begins orally and via designated vessles to send it greater nourishment; and it grows. Yet, it remains most tender and soft wrapped within its membranes and is still called an embryo. Having acquired the necessary vigor and revealed all of its parts it is then called a Fetus and it prepares for release.[55]

For Vallisneri, eggs, not spermatic worms, were therefore the main principle of all generation.[56]

It is necessary to point out, however, that while each of these authors emphasized the particular and indispensable organs and functions of the female body for human reproduction, each also adhered in varying themes and degrees to ancient concepts of female deficiency and male supremacy. De Graaf's strikingly expressive, often anecdotal account of the female reproductive system offers a colorful recapitulation of certain leitmotifs of the misogynist anatomical tradition, as when he describes the insatiable uterus with hymen intact of a pregnant fourteen-year old girl: "When semen poured in between the lips of her pudendum, her uterus, greedy for this kind of food, drew it to itself, just as a stag attracts snakes out of deep holes with the breath of its nostrils."[57] In a series of lessons on human reproduction given at the University of Padua from 1711 to 1715, Morgagni explicitly reiterates a core Aristotelian axiom that the female contributes only the material cause of the fetus, while the male contributes the nobler efficient, formal, and final causes:[58] "we observe that our theory admirably agrees with the hypothesis of Aristotle: the material of the fetus comes entirely from the mother . . . while exclusively from the paternal contribution obtains the principles of motion and energy."[59] And as regards Malpighi, Catherine Wilson summarizes that he believed "that females in general produce eggs and cast them off, and that these eggs, when fecundated by the male 'unfold into a new life.'"[60] Malpighi's notions of human conception were, moreover, inextricably linked to his vast knowledge of plant generation. In his analysis of the cause of sterility in a Luchese couple, he states:

> The generation of man is so obscure that it can be explained only by analogy with perfect animals and with plants. This is certain: Nature has constituted twin principles, an active and a passive, the latter producing the egg, which the other principle fecundates by moistening it with semen. Moreover, in order that [the egg] may grow like a plant, the semen must be committed to the uterus as to its proper soil, and when favorable colliquament emanates from the uterus, the parts of the animal become visible, grow, and are made strong.[61]

For his part, Vallisneri was a well-known friend and admirer of learned women of his age, notwithstanding his judgment in the famous debate on women's education waged by the Paduan Academy of the Ricovrati in 1723, in which he ruled that formal instruction be limited to noblewomen. (His ruling in the debate did, in fact, provoke widespread outrage among women in and beyond Padua and generated formal rebuttals by women writers even thirty years after the event.)[62] In his scientific writings on human generation, ancient contrasts between the sexes clearly if unsystematically resurface, as in his comparison of the superiority of male to female spermatic worms:

> There are worms of two types, that is male and female, such that if a male enters the egg, a male will be born, if female, a female will come to light. All merit and fortune depends on the greater or lesser vigor of the worms. . . . Thus it is that we see that in as much as man is more robust, they will be born male because they are better nourished and more active. But if [the worm] is weak and consumed, a female will be born.[63]

Anna Morandi's representation of the reproductive male body not only overcame dominant masculinist theories of the reproductive body and sexual difference, but also rejected the long-established *anatomical rivalry*: male versus female. She conceptualized the reproductive male body outside the male/female binary while at the same time attributing to it a reproductive imperative parallel to that of the female body. By focusing on the specific generative functions of the components comprising the male urogenital system, Morandi implicitly rejected difference as a requisite means for delineating the processes and final cause of human reproduction. As will be shown, she looked at the male anatomy in isolation, indexing its component parts and their distinct and collective form and function for human reproduction.

ANATOMICAL PREPARATION OF THE MALE GENERATIVE PARTS CONSTRUCTED BY ANNA MORANDI MANZOLINI, BOLOGNESE CITIZEN, ANATOMIST, AND HONORARY ACADEMICIAN IN THE INSTITUTE OF SCIENCES OF BOLOGNA[64]

With this weighty heading in her *Anatomical Notebook* identifying the subject of Morandi's investigation and her notable academic authority to perform such a study, she begins analysis of the parts of the male reproductive system. As in all her anatomical studies, Morandi first exhibits the intact, animate body, in this case "the trunk of a man in its natural situation in which the integuments [coverings/cuticle/membranes] in the lower abdomen are opened and all the viscera removed that would impede a clear vision of the generative parts of the male sex."[65] She proceeds methodically, as in dissection, from the uppermost parts of the specimen to the lowest, and from exterior to interior. It is important to note, however, that in contrast to today's method of regional dissection, according to which the abdomen is typically divided into quadrants or nine regions "by imaginary vertical and horizontal planes through the umbilicus," Morandi viewed the anatomical body systemically.[66] In Table One, she renders a macrocosmic view of the facing retroperitoneal space and pelvic cavity from which will methodically follow her anatomies of the interconnected parts of this system and the discrete, ever smaller, even microscopic, details of its composite form:

Integument
Diaphragm
Portion of the Esophagus
Descending aorta
Inferior Vena Cava
Emulgent vessels[67]
Adipose vein[68]
Kidneys
Supra renal capsule
Spermatic vessels
Ureters
Psoas muscles
Bladder surrounded by peritoneum
Transversal or semicircular fold of peritoneum, or suspensory ligament of the
    bladder [medium umbilical fold or ligament]
Urachus
Portion of rectum
Part of large intestine
[Page 97r]
Vas deferens
Holes of the aponeuroses of the oblique muscles of the lower abdominal ingui-
    nal rings, through which pass the spermatic vessels that go to the testicles
    and through which also pass the vas deferens that comes from the testicles
    going to the seminal vesicles.
Portion of the pyramidalis muscle
Penis

Glans Penis

Prepuce

Scrotum that enclose the two testicles

Median Raphé, which is like a seam that keeps the two testicles divided and
    separated, the one from the other. Closed inside what anatomists have called
    a purse, the testicles are destined by nature to remain outside of the lower
    abdomen for two reasons: the first is to modify the heat of the sperm or
    semen; the second, to maintain the tautness of their conduits by means of
    gravity.[69]

Fat covering the pubic bone

Inguinal region

Two truncated thighs[70]

In accordance with the canon of modern, empirical anatomical design, which she must also be seen as helping to elaborate, Anna Morandi's literal overview of the subject, from diaphragm to severed thighs, indexes the primary anatomical components of the urogenital system. Atlases and anatomical studies by Cowper/Bidloo, Vesalius, Vesling, Giulio Casserio, Morgagni, and numerous others that filled the shelves of her private archive served at most as archetypes and more narrowly as points of departure for her anatomical studies. Just as Morandi quoted illustrations in Cowper/Bidloo of intact, looking eyes and the branching ocular muscles, she doubtless modeled her sculptures of the male body to a limited extent on canonical illustrations (figs. 65–66).[71] However, her work was devoid of traditional traits of the anatomical atlas that Andrea Carlino has aptly described as literary and typographical ornament: heroic landscapes, and iconographic figures laden with symbol and metaphor.[72] Indeed, as this first figure helps to show, Morandi's method of anatomical demonstration sharply deviated from that prescribed to artists by Leon Battista Alberti.[73] While the artist reconstructed from bone to muscle to skin the machinery of the body in motion for verisimilar and aesthetic display, Morandi mined the depths and minutiae of the living body step by step in order to distinguish the origins and extension of discrete anatomical parts and systems, their physiology and interconnections.

Nevertheless, in the absence of her actual wax sculptures of the male reproductive system, nearly all of which are lost, it is possible to approximate what the first figure in her series looked like based on discrete anatomical images well known to her. Andreas Vesalius's *Fabrica*, whose illustrations were imitated in the works of Antonio Maria Valsalva, Casserio, and others, likely served as a principal font for this facsimile. In her study of the male reproductive system, Anna Morandi, like Vesalius, began with the trunk of a man, with organs of the lower abdomen exposed (fig. 67). Although her figure would almost certainly have exhibited a less heroic posture and musculature than

FIGURE 65. Anna Morandi, eyes, wax. Courtesy of Museo di Palazzo Poggi, Università di Bologna.

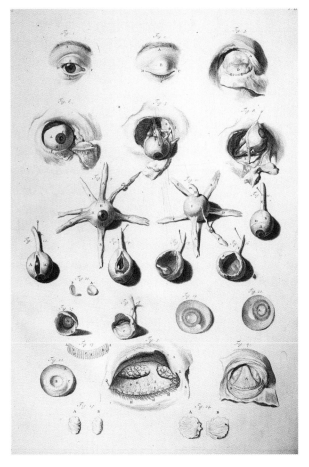

FIGURE 66. William Cowper, table 11, in *Anatomia corporum humanorum* (1750). Courtesy of Becker Medical Library, Washington University School of Medicine.

the Vesalian archetype, based as it was on the torso Belvedere, the general content probably resembled that of her precursor. Apart from Vesalius's inclusion of the liver, which Morandi extracted, the mode and general syntax of each author's expression of the male urogenital system are broadly analogous, having been based in both cases on dissection.[74]

Following this overview, Morandi subsequently isolates and details the key parts of the system, beginning with the passage of the urethra from the bladder to the tip of the glans penis. This is followed in order by figures demonstrating

the anterior and posterior of the bladder opened to show the passage of the urethra;

the prostate and the seminal collicles [more commonly known as the verumontanum];

the penis with and without its coverings;

the two spongy bodies of the corpus cavernosum that make up the penis;

the seminal vesicles, their parts and secretions; the seminal vesicles, the prostate, and the antiprostate [now known as Cowper's gland];

the bladder distended with urine;

the prostate shown to expose the passage of the bladder to the urethra;

the scrotum shown with the testicles in and ex situ;

various figures and sections of the testicles; a macerated testicle to show the silken threads of which it is composed;

the integuments of the scrotum, testicles, and penis;

the kidneys and their substances and papillae radiating in the pelvis;

the sectioned kidney;

the tiny blood vessels of the testicle shown by means of wax injections;

the arterial, venal, and nervous vessels that nourish the genital parts;

and finally

the pelvic bones.

Her visual and literary account thus simultaneously strives toward deeper and deeper views, as well as the particular, the detail, of ever more minute, even microscopic, components and substances within the larger reproductive system. Although no documents exist that detail the extent and method of her work with the microscope and such techniques as wax injection, the use of this technology was by this time routine in centers for the study and teaching of empirical science and anatomy.[75] As this summary suggests, in fact, Morandi's bipartite visual and written study of the male urogenital system was fundamentally didactic. The anatomist methodically schooled her student-spectators in the structure, organization, and function of the reproductive system on display.

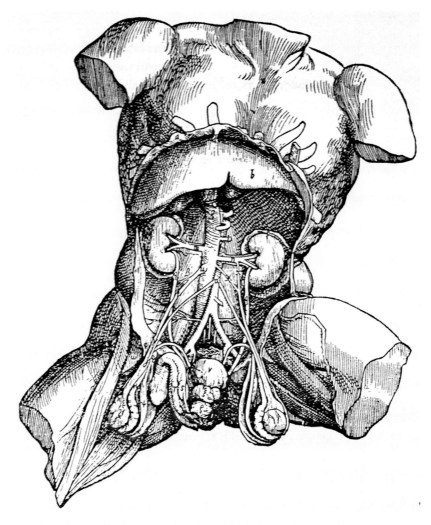

FIGURE 67. Andreas Vesalius, fig. 22, in *De humani corporis fabrica,* book 5. Courtesy of the Rare Book Archive of the Bernard Becker Memorial Library, Washington University.

Her analysis excels in clarity, precision, and completeness of detail. This is especially clear in Table Four, in which she demonstrates the structures involved for the passage of semen from the testicles to the urethra. Her straightforward description again emphasizes what was seen and from which precise perspective during the course of dissection:

In this is shown two bladders in their natural situation but opened, one from the posterior, the other from the anterior; in this the prostate gland is opened in order to show the seminal collicles [*grano ordeaceo,* or "barleylike grain"], still called verumontanum. Along with the prostate gland, the urethra is opened to the end, by which one can see the passage of that canal that serves both the semen and the urine. Aside from the sphincter of the bladder, this shows the opening of the urethra inside the bladder after having passed obliquely through the three membranes.[76]

Anna Morandi no doubt explained the components of the male urogenital system and their functions to the medical students, dilettantes, and tourists who attended her studio lessons. Her notes convey the didactic voice with which she likely addressed her public, as when she elucidates the parts of the bladder that enable the retention and expulsion of urine:

> The substance of the urinary bladder is composed of three membranes. The first is called *carnosa* [fleshy, muscular], the second *nervous*, and the third *villous*. The first membrane of the bladder is composed of three strata of variously woven fibers, the first marked by the symbol *E*. These are generally the straightest and are destined to attract the urine; the second, marked by *FF*, are oblique, and have the function of casting out the urine, the last marked *GGG*, are crossed at a sign near the neck of the bladder where they come to form an entire sphincter, which one can see in this figure, and these are destined by reason of their circumvolution, or sphincterlike form, to hold the urine in the bladder so that it does not escape without our consent, just as Galen and Vesalius thought.[77]

As Nancy Siraisi has observed, inherent to the tradition of modern dissection was not only the itemization and analysis of the anatomical subject, the cadaver or body part, but the complementary dissection of the accuracy and relevance of master anatomical texts.[78] Adhering to this tradition, here Morandi confirms the theories of Galen and Vesalius, while implicitly, but unmistakably, surpassing her forebears' scientific precision.

The Purposive Design of the Male Anatomy

Beyond its didactic design, evident in Morandi's study of the male anatomy are essential teleological concerns. She adheres in this and select parts of her oeuvre to a tradition in anatomical science most explicitly articulated by Galen in his work *On the Usefulness of the Parts of the Body*, which presumes the perfect utility and constitution of all parts and aspects of the body for realizing the functions of the soul as designated by the divine craftsman—or what he terms, following Plato, the *demiurgos*.[79] Andreas Vesalius followed Galen closely in his advocacy of a teleological conceptualization of human anatomy and physiology, making explicit reference to their ideal design by that divine force he variably identified as the "Opifex," the "Founder of Things," the "Creator," and "God,"[80] while "Nature," for Vesalius, denoted God's handmaiden and intermediary in the physical world. As Siraisi points out, however, for Vesalius scientific accuracy in explicating the parts and functions of the body represented the supreme mode of venerating the original Anatomical Designer.

setting despite their being of such incredible subtlety that in all of their total-
ity they form a very tenuous net even while also demonstrating the form of
the testicles themselves."[93]

The intensified attention to purposive design is accompanied in Morandi's
notes by an increasingly suggestive commentary on the process of male fecundity, culminating in her description of the seminal vesicles, where she mistak-
enly believed semen collected as in an antechamber for finishing enrichment
before ejaculation:[94] "the seminal vesicles are exposed, which are a reservoir
for that balsamic liquor called semen." The reference to semen as a woody
fluid connotes soothing, healthful, and vital properties while linking the fertile
male body to the life force that animates other parts of the natural world.[95]
In the same figure and in terms reminiscent of the French anatomist Jacques
Bénigne Winslow, whose work she studied, Morandi describes opening the
seminal vesicles "in order to show those velvety or glandulous coves or chambers where the semen resides and continually perfects itself."[96] And it is here
within the downy hollows of the vesicles that she discovers "a certain particular
juice . . . that ever refines and perfects the semen"[97] by restraining its inherently
volatile generative force. A leitmotif of Morandi's theory of male fecundity,
she held that an accretion of substances from the testicles, the vesicles, and the
prostate combine with the semen en route to the urethra to enrich and steady
its impulsive generative capacity. Notably, notwithstanding certain minor
inaccuracies and poetic excess, on this subject she was not wrong.[98]

Anna Morandi's repeated use of the term *perfection* denotes completion,
according to Latinate usage. Of course the notion that semen undergoes a
process of perfection or completion as it is formed along the paths of the
urogenital system was hardly new with her account. Indeed, theories on the
process are pervasive in ancient and early modern anatomical texts.[99] With
her more modern counterparts, she viewed the testicles as the prime site for
the production of semen. However, she theorized that the whole urogenital
system contributes to the full realization of the generative potency of the
semen, which takes place serially as the semen travels the numerous path-
ways leading to the urethra, accumulating critical nutritive and congealing
components along the way, until the final stage of ripeness transpires in the
seminal vesicles.[100] Morandi's systemic view of male fecundity differs markedly from the theories of such authorities as Regnier de Graaf and Giovanni
Battista Morgagni, which revised ancient notions of male generative potency.
Supplanting pneuma, Aristotle's "vector of the soul," with "animal spirits"
that the nerves carry to the testicles to mix there with blood conveyed by the
spermatic vessels, Morgagni and de Graaf offer a new recipe and lexicon for
seminal confection into what in ancient times Hippocrates called the "effer-

vescent foam of life."[101] Morandi's very different emphasis on the process of becoming, of completing, of satisfying a reproductive imperative makes no reference to a point on the trajectory of male fecundity when seminal fluid is infused with an ennobling spirit, or life force. Her account may, in fact, be seen as a democratizing gesture through the imputation of a traditionally feminine principle to the male body. Morandi portrays the reproductive male body, in ways that parallel traditional conceptions of female anatomy, as immanent, tending toward completion through an obligatory, though involuntary, performance of nature's ideal generative design. She defines the male body by its interior and, most important, its all-encompassing reproductive function.

The most distinctive aspect of Morandi's teleology of the male reproductive body is, however, the absence of a feminine foil. No corollary account of female difference or deficiency attends her explanation of the process of seminal completion and male fecundity.[102] Nor does Morandi assert, with many of her precursors, "the pre-eminence of the product formed by the male" for reproduction.[103] The male reproductive body is, for her, ideally designed as a composite whole and in all of its parts for the generation of the species. She does not identify a predominant organ for male reproduction or a material point of origin for male ontological difference. She offers no "thinking penis" as complement or counterargument to Petronio Zecchini's "thinking uterus." At the same time, she ascribes to the male body a reproductive imperative both essential and integral. From fascia and membranes to arteries, veins, and vessels; kidneys and bladder to urine, blood, and chyle; integuments to penis, testicles, and pubic bone, all parts serve the progressive processes of fertility and reproduction. Indeed, the symmetry of the male reproductive imperative with that of the female suggests a kind of teleological balance of the sexes.

As has been noted previously, Morandi's anatomy of the reproductive female body is without written gloss and limited in scope primarily to her patron's commission. However, given the hundreds of extant wax figures and pages of scientific notes on myriad parts of the body by Morandi, which steadfastly manifest her aim for maximum scientific accuracy as well as her consistent conceptualization of bodily components within the context of a vital, interdependent anatomical and physiological whole, it is fair to presume that her extensively elaborated poetics of anatomy would not have dissolved before the female body. Nor would Morandi, who in her written analyses boldly interrogated and often disputed the theories of such canonical authors as Galen, Vesalius, Valsalva, Malpighi, Rivinus, Winslow, and Adriaan van den Spiegel, have likely yielded to the misogynist tradition of representing the female anatomy. Indeed, the few occasional waxes of the female reproductive system that form part of her private collection are consistent with her written descriptions of her wax renderings of the reproductive male anatomy. Her illustration of the uterus and its realistic dimensions after labor and delivery shows the vital organ open, round, and distended with ligaments, muscle, and fascia attached. The expression of the tissues adjoining the uterus alludes to the larger context of the lower abdomen, just as Morandi's representation of a rippling swirl of placenta that seemingly spills over the side of the wooden display table in her facsimile of the anatomy of a newborn child indicates the origin and furious motion of that birth (figs. 68–69).

In the actual and ideal anatomical theater she occupied, Anna Morandi asserts her authority through neither a reiteration nor a reversal of the sexual imbalance. As with all the parts of the body under her scalpel and her gaze (fig. 70), she seeks to demonstrate what is dispassionately revealed by experience in accordance with the modern method, while at the same time acknowledging the perfection of nature's complex design.

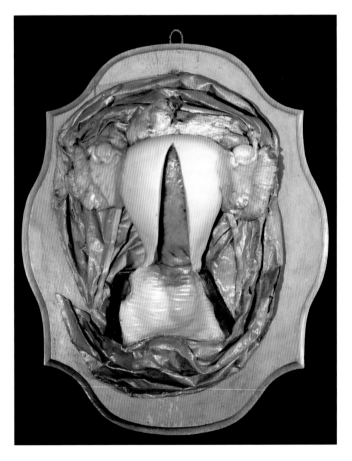

FIGURE 68. Anna Morandi, obstetrical figure. Courtesy of Museo di Palazzo Poggi, Università di Bologna.

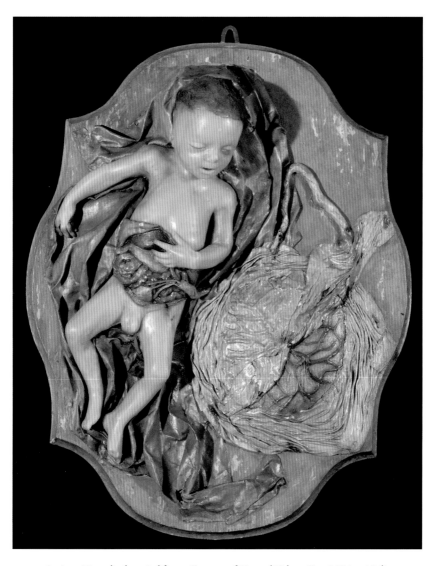

FIGURE 69. Anna Morandi, obstetrical figure. Courtesy of Museo di Palazzo Poggi, Università di Bologna.

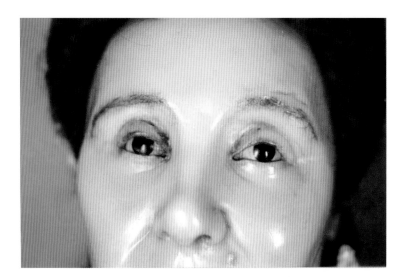

FIGURE 70. Anna Morandi, self-portrait (fig. 41), wax; detail. Museo di Palazzo Poggi, Università di Bologna. Photo by author.

7

Cessio ac Venditio

The Final Years and the End of an Age

Bertalia, Italy
3 September 1765

Most Gentle Signore,

Your most lowly servant Anna Manzolini greatly humbles herself
to Signore Marcello Oretti and at the same time beseeches him to
graciously intervene as they had arranged together regarding her desire
to request from the Most Illustrious Signori of the Congregation for the
Administration of Customs [Congregazione del Governo della Dogana]
for an increase in her honorarium, established some time ago. [This
subvention would serve] her need for 200 more lire annually. The reason
is that she finds herself in great want because of a grave illness and because
she was not given the support she needed to survive by those who should
do their duty. She writes with all due respect to plead again.

    ANNA MORANDI MANZOLINI, letter to Marcello Oretti

From her residence in the northwest Bolognese countryside of Bertalia, a
highly distressed Anna Morandi wrote several versions of the above letter[1]
in a nearly illegible hand to implore the immediate aid of Senator Marcello
Oretti. Despite Oretti's avid support and his intervention on her behalf, the
Bolognese Senate refused her request for the modest 200-*lire* increase in her
honorarium. The rebuff finally obliged Morandi to undertake seriously the
transfer of her practice and the sale of her anatomical wax collection to one
of the European academies and states extending offers to her.

    It was well known among Bolognese civic and cultural leaders that, for
some time, Empress Catherine the Great had eagerly desired to bring the Lady
Anatomist and her celebrated collection to the Russian court. From shortly

after the death of Morandi's husband ten years earlier, in 1755, the empress had extended a standing invitation to the Signora Manzolini.[2] Confidential negotiations now commenced for this move. Despite the inadequate support she faced in her own city, however, a permanent relocation to a distant foreign capital was a last resort for the Bolognese anatomist, who, especially in her growing infirmity, wished to remain at home. There is, in fact, no record of her ever having traveled beyond Bolognese territory to any of the distant or nearby parties that had formally solicited her with invitations, the promise of academic posts, and even a blank check.[3] Therefore, when a local remedy to her financial ills arose in the midst of her negotiations with the empress, she immediately took it. As Oretti writes in his biographical sketch of her, Signora Manzolini "was forced to search elsewhere for support and so took advantage of the opportunity extended to her by Senator Girolamo Ranuzzi."[4]

Count Ranuzzi (1724–1784), a Bolognese nobleman and masterly huckster, who cultivated for his personal gain relationships with prominent individuals throughout Italy and Europe, saw Anna Morandi as another means for profit and prestige. He thought to add wax anatomical figures to his other entrepreneurial ventures, which included the sale and distribution of mineral water from his land in Porretta, and traffic in such commodities as cocoa, tobacco, horses, and even the architectural designs of his palace.[5] According to the notarized bill of sale of Morandi's complete collection of wax anatomical figures to Count Ranuzzi on 6 May 1769:

> After the news reached the Excellent Lord Senator Count Girolamo Ranuzzi that Signora Anna Manzolini was in a position to sell the anatomical preparations that she had sculpted in wax, and that she was in the process of negotiating the sale of her oeuvre to a Foreign Person, the said Lord Senator, desiring to have such a renowned and prestigious collection for himself, petitioned Signora Anna Manzolini to permit his acquisition of it, to which request she assented for the sum of 12,000 lire . . . [Count Ranuzzi was further] obliged to grant and assign to said Signora Anna Manzolini . . . for the duration of her natural life, an unfurnished apartment [in his palace] with all those necessary comforts for her happiness.[6]

Ranuzzi thus acquired the entire collection of anatomical wax figures mounted to nearly one hundred display tables by the famed Lady Anatomist. In addition, he obtained the exquisite busts Morandi sculpted in wax of her husband, Giovanni Manzolini, and herself; one upright full-size skeleton, and another of a two-year-old with its attached ligaments; and four large armoires "finished with their necessary fastenings and glass" to exhibit with elegance the celebrated oeuvre.[7] Ranuzzi eventually purchased on 6 April 1771—for the modest amount of "lire 600 quattrini," to be paid in 100-*lire* installments over six years—Morandi's library of some twenty master anatomical atlases and surgical texts, her dissecting and sculpting instruments housed in two armoires, and the wax bust she had created of the late Father Ercole Maria Isolani (1786–1756), of the Oratoria in Bologna.[8] A mere eight months after her death in 1774, the count wrote to the Bolognese Senate, offering to sell the complete collection. He received the 16,000-*lire* payment in full from the Senate in 1776, a respectable 20-percent profit over his original investment.[9]

Morandi's residence in the Ranuzzi palace with her twenty-one-year-old son, Carlo, and thirty-five-year-old maidservant, Lucia Grifoni,[10] proved an arrangement mutually beneficial for boarder and host. The massive Palladian edifice to the south of the city center, which is today Bologna's Palace of Justice, was a chief stop for eighteenth-century Grand Tourists. The architectural features, grand double staircase, tapestries, frescos by Domenichino and Bigari, and notable art collection that included works by Raphael, Titian, Tintoretto, Guercino, Annibale Carracci, Reni, and Albani were well known to foreign visitors by their repeated listing in the published travelogues of the day. After her own installation in a large, elegant upper apartment, Anna Morandi, too, became a principal and published attraction for the many visitors to the palace. Thus, after years toiling alone against financial adversity and the mental and physical strain that attends such troubles, Morandi could finally live in security and comfort and advance the celebrity of her practice.

Girolamo Ranuzzi was, in fact, a tireless promoter of her celebrity, and thereby his own. He hosted the visits of numerous eminent guests to her studio and mediated Morandi's affairs with her foreign patrons. Letters to Ranuzzi from the Baron de Staal of St. Petersburg and Maximilian Frederick, Elector of Cologne,[11] record their orders for anatomical wax figures.[12] When it served his financial interests, however, Ranuzzi also occasionally interfered with his houseguest's private transactions, as when he discovered that Morandi had received a commission from Princess Jablonowska of Breslau to create a series of the sense organs. The count asserted his rights over these waxworks as proprietor of Morandi's complete oeuvre.[13] Moreover, Ranuzzi sought to broker the sale of a series of Morandi's models to Empress Catherine the Great in 1770, as indicated in a recently discovered letter of 1776 he sent to the Russian sovereign. This letter will be discussed subsequently at length.

Most spectacularly, on the morning of Pentecost Sunday, 14 May 1769, immediately after Anna Morandi's move to the count's palace, Ranuzzi hosted, with all the necessary honors and comforts, the two-hour visit paid to her and her collection by Emperor Joseph II of Austria.[14] Ranuzzi meticulously documented the cost to him of the visit, eventually calculating it into his offering price to the Senate for the sale of Morandi's oeuvre.

During the encounter, the emperor, who had previously expressed an avid desire to meet the Lady Anatomist, inspected Morandi's collection of wax anatomies and interviewed her at length about her practice and her ability to undertake the gruesome work of human dissection. As a sign of his esteem, he presented her with a gold medallion bearing his image on one side and on the other a globe and the helm of a ship encircled by a laurel branch and the motto *Virtute et Exemplo*.[15] Historians have attributed Joseph II's inspiration for the 1784–86 commission and installation of more than one thousand anatomical and obstetrical wax models in the Josephinum, his Military Medical Academy in Vienna, solely to his 1780 visit to the wax anatomy collection that his brother, Grand Duke Peter Leopold of Tuscany, had commissioned for Florence's Specola Museum of Natural History. It is fair to presume, however, that the initial impulse to obtain anatomical wax models for the purpose of practical medical instruction awakened in the emperor during his visit with Anna Morandi. It was here that he would have first seen the division of wax anatomical models into separate bodily themes, displayed in designated cases. The evolution of Morandi's collection, as has been noted previously, adhered to the systematic disclosure of the body in dissection as well as to canonical representations in anatomical atlases. The structure and organization of the Josephinum collection of wax anatomies reiterated on a vast scale this same organization, also replicated in the Florentine Specola. Indeed, Felice Fontana, the natural philosopher and court physician charged late in the century by Peter Leopold with the development of the Florentine anatomy museum, had himself taken his inspiration from the work of Morandi, Manzolini, and Ercole Lelli during his years as an academician in Bologna.[16]

One year after Joseph II's visit to his palace, Girolamo Ranuzzi commissioned the famous English neoclassical sculptor Joseph Nollekens (1737–1823), sojourning in Bologna on his way back to London, to sculpt a bust of the Lady Anatomist for 302 *lire*. The bust was unmistakably another way to garner public notice for his illustrious tenant and, with its sale to the right person, a nice profit for himself.[17]

Known for the realism of his sculpture portraits, Nollekens's depiction of Morandi is a dramatic departure from her idealized self-portrait in wax (fig. 41), as we have seen, showing her in ornate aristocratic attire in the act of dissecting a human brain. In the Englishman's vision of her, she appears a rather unremarkable elderly woman, shown to the chest and turned three quarters, looking impassively over her left shoulder. Devoid of all marks of wealth and high social station, her dress appears quite plain and humble, as does her hairstyle, pulled up under a simple cap. Significantly, there is no sign of her profession—no scalpel, forceps, or sected body parts. Indeed, as Nollekens portrays her, she is altogether without the instruments of her arms and hands.

Two years after Morandi's death in 1776, Ranuzzi sent a bronzed plaster cast of the bust to Catherine the Great to "perfect" the collection of wax anatomies the empress had previously commissioned from Morandi in 1770 by way of the count. Ranuzzi's letter to Catherine heralding the arrival in Russia of the bust[18] offers crucial new proof of the sovereign's abiding fascination with Anna Morandi and her wax anatomies, as well as of the count's capacity for creative merchandising and self-promotion on an international—indeed, an imperial—scale:

> Your Majesty,
>
> All that is a tribute to the fair sex is homage owed a Princess, whose virtues and enlightenment honor the Throne and Mankind. For this I was so bold in the year 1771 as to offer Your Imperial Majesty the bust modeled from life of Lady Manzolini, whose collection of anatomical tables your Majesty deigned to acquire. In the year 1770, I received that Famous Woman in my home in order to guarantee her a tranquil setting. She regretted that old age and the infirmities that accompany it had not allowed her to obey the orders of Your Imperial Majesty [to move her practice to Russia]. She had hoped to live a long time in my home and in her own climes, but she had the misfortune of contracting a disease from which she died last year. In consequence of her death and in order to enhance the aforementioned collection of anatomical tables, if Your Imperial Majesty will permit me, it would be my honor to dispatch that same bust of Lady Manzolini to your Majesty. You reign, Madame, over the vastest empire in the universe, and your Soul is even greater than your states. For this reason, I beg Your Imperial Majesty to forgive the boldness which I hope to take and which I, in fact, took, when I consigned to Mr. De Grimm, plenipotentiary minister of France, a book of prints of my palace, which is considered a curiosity in my country, asking him to present it in my name to Your Imperial Majesty, with the homage of a foreigner who wishes Your Imperial Majesty a long life. It is with similar sentiments that I am, with the utmost respect of Your Imperial Majesty, the very humble and obedient servant.
>
> Count Ranuzzi,
> Senator of Bologna.
> Bologna, Italy, 6 May, 1776.[19]

Three months later, in a letter of 4 August 1776 to the French minister and philosophe De Grimm, Catherine delightedly tells of her receipt of the bust and her prominent display of it on her dining table at her summer palace in Peterhof. Together with Girolamo Ranuzzi's correspondence, this letter crystallizes Catherine's avid interest in the life and work of Anna Morandi and her mischievous inclination to affiliate herself personally with the Lady Anatomist:

> With regard to the portrait or bust of the famous artist from Bologna, I keep it on my table at my palace in Peterhof, and everyone asks me: Who is this? And I, in order to dismay tiresome interrogators say: This is my grandmother.[20]

Regrettably, the bust was destroyed during World War II. Its existence can be traced in custodial records at Peterhof until 1938, at which time it was still displayed on a specially made pedestal in the "Sofa room in the women's half of the palace." Several extant photographs attest to its existence until that time.[21]

Beyond the few notarized documents recording Morandi's business dealings with Ranuzzi, church records of her residence in the palace, and official descriptions of Joseph II's visit, there is scarce information about how she lived her final years until her death on July 9, 1774. Commissions from foreign patrons reveal that she maintained her anatomical wax-modeling practice through the end of her life and continued to seek international attention at the highest levels. A letter sent to her from Vienna on 17 December 1770 from the foremost librettist of the century, Pietro Metastasio, indicates that she had sought his aid in promoting her anatomical waxworks within the imperial court, where he was the longtime poet laureate, and confirms her unremitting endeavor to secure top international patrons for her work.[22] Although it is possible that she produced new models during this time, it is more likely that she recast from plaster molds copies of previous work, such as her series of the sense organs that she was completing for the princess of Breslau when she died. It is highly unlikely that she continued to dissect cadavers and teach anatomy to medical students after her move to the Ranuzzi palace. Her quarters, opened to the public, served essentially as a salon for receiving notable visitors and as a museum for the display of her anatomical collection. Ranuzzi, moreover, would likely have been loath to see to the particular and more objectionable resources for a working anatomy laboratory in his palace, including a dissection table and a steady flow of fresh body parts from the city morgue. The very instruments she used for dissection had become by 1771 another of Ranuzzi's investments, and lay neatly and stone still alongside the rest of her exhibited works inside their large glass cases.

The end of Morandi's life is illumined to an extent by the inheritance she left to her two sons after she died. She had arrived in the Ranuzzi palace on the brink of financial ruin in 1769; her bequest little more than four years later consisted almost entirely of the remaining payments owed by Girolamo Ranuzzi for the purchase of her oeuvre. Morandi's last will and testament unevenly apportions her estate between her aristocratic eldest, who had been adopted fifteen years prior from a Bolognese orphanage by the noble Solimei family, and her dependent youngest. It also hints at her asymmetrical division of maternal feeling between the son she kept and the son she gave up. On 18 July 1774, the notary Domenico Schiassi opened and read before the thirty-one-year-old Giuseppe Solimei and the twenty-five-year-old Carlo Manzolini the following will left by their mother:

> The Will of me, Anna Morandi Manzolini, consigned to the Notary Signore Domenico Schiassi that I signed with my regular signature this day 24 April 1770. I, the undersigned Anna, daughter of the late Signore Carlo Morandi, widow of the late Signore Giovanni Manzolini, while by Divine Mercy, I am sound enough of mind, of all other senses, and also of bodily health to dispose of my earthly belongings, I make the following arrangements, that is, with my soul separated from my body, I would like it to be entombed in the Church of the Parish in which I died. To Lucia Grifoni, my servant, I leave the amount of fifty lire. To Signore Giuseppe Solimei, born Manzolini, my legitimate first son . . . I leave all that by law should come to him as a portion of my assets . . . the corresponding percentage of Credit that perhaps at the time of my death I will have from the Noble and most Excellent Count and Senator Girolamo Ranuzzi as residual earnings for the sale of my Anatomical Preparations, completed on 6 May 1769. . . . I also leave to Signore Giuseppe the gold medal with the portrait of His Majesty Joseph II, Emperor, given to me as a gift by His Majesty during the visit that he made to this City. . . . Not for any malevolence I hold against him, for whom I preserve affection and benevolence, but rather being Signor Giuseppe abundantly provided with Wealth due to his heredity as a Solimei . . . [while] my other son will not be provided with any other than the effects of my estate, . . . [therefore] of my Universal inheritance, I consider, call, name and want that my Universal heir be the Deacon Carlo Manzolini, my very beloved [*dilettissimo*] youngest son presently enrolled as a student in the College Seminary of this city.[23]

Anna Morandi was buried in the local Church of San Procolo, to which the Ranuzzi family belonged. The enormous marble headstone purchased by her two sons that prominently marked her grave in the floor of the nave finally made rock-firm the glory of her singular achievements so often devalued by her Bolognese compatriots during her lifetime. In great block lettering, her Latin epitaph acclaimed:

Annae Manzolinae	Anna Mansolini
In Patrio Gymnasio anatomicae	Anatomist of her Native University
In Florentissimas Italiea Academias Cooptata	Member of the Distinguished Academies of Italy
Amplificatrici Facultatis Suae	Enhancer of her Own Discipline
In Fingendis e Cera Humani	In Shaping Parts of the Body from Wax

Corporis Paritbus	
Supra Omnes Retro Artifices	Superior to all Previous Artisans
Praestantissimae	Most Outstanding
In Iisdem Explicandis disertissimae	Most Learned in Explaining the Same
Tanta Celebritate Famae Ut Eam	Of Such Great Fame that
Josephus II Augustus	Emperor Joseph II
Adierit	Visited Her
Tanta In Patriam Charitate	Possessed of such Love for her Homeland
Ut conditionibus amplissimis	That the most generous offers
Saepe Repudiatis	She often Rejected
Civium Suorum Causa	For the Sake of her Fellow Citizens
In Mediolanensium Londinensium	To the City of London
Petropolitanorum Accademias	To Sculpture Academies
Ultro Arcessita	Summoned without Having Asked
Venire Nolverit	She was Unwilling to Go
Quae visit Ann LVII	Who Lived Fifty-Seven Years
Obit VII Id Jul Ann MDCCLXXIV	Died 9 July 1774
Josephus Solimeius A Calolus Manolinus	Giuseppe Solimei and Carlo Manzolini
Filii	Her Sons
Matri Rarissimae Incomparabile Benemerenti	For Their Most Singular and Incomparably Deserving Mother
Maestissimi Posuerunt	Have Erected This Most Sadly[24]

This same tombstone that has lain hidden in the church's crypt since 21 March 1873, when Morandi's remains were disinterred and moved to the Certosa Cemetery outside the city's walls, now serves as a sign and symbol of her obscurity in the annals of Bologna and the history of eighteenth-century art and medical science.[25]

Morandi's son Carlo was allowed to remain in the Ranuzzi palace for one year following his mother's death. He had taken orders in 1770 to become a deacon of the Basilica of San Petronio, and continued in these studies after his departure from the palace.[26] On 3 April 1781, the Bolognese Senate appointed him professor of theology at the university. He rose to the prestigious post of canon of San Petronio the following year. In 1814, under the authority of the Kingdom of Italy, he was appointed professor of ecclesiastical history. He died in 1826 at the age of seventy-seven. Records are lacking for the fate of Morandi's eldest son, Giuseppe, and his noble inheritance.

Count Ranuzzi initiated negotiations to sell Anna Morandi's oeuvre immediately upon her death. To his strategic advantage, a "Foreign Person," undoubtedly the ever-looming Russian sovereign, again made known her desire to buy the collection. Playing on the threat of desire from abroad for Bologna's matchless waxworks, ardor practically substantiated, he claimed, by the outsider's willingness to pay the full value of 7,000 *scudi romani* for the oeuvre, Ranuzzi offered up the collection to the people of Bologna for a *mere* 6,000 *scudi*. He professed to his fellow senators his patriotic aspiration to prevent great injury to the "public well-being" and, more important, to Bologna's cultural patrimony should Morandi's marvelous sculptures be carried away to a strange land. Members of the Bolognese Senate reacted at best with restrained interest in the proposal and at worst with umbrage at Ranuzzi's "brazen and excessive" selling price.[27] The count was undaunted, however, and with characteristic tenacity eventually persuaded Bologna's civic leaders to buy him out.

On 25 June 1776, Count Girolamo Ranuzzi and Giuseppe Angelelli, president of the Institute of Sciences, reached an agreement on a price of 16,000 *lire*, approximately 40 percent of Ranuzzi's original asking price, for Morandi's anatomy studio, to be paid in installments over four years. Included in the sale was everything still on display in the Lady Anatomist's apartment in the Ranuzzi palace: her complete collection of waxworks, anatomy texts, and surgical instruments; a series of skeletons from fetal stage to adulthood; the wax portraits she sculpted of her husband, Father Isolani, and herself; the large armoires exhibiting all the above; Nolleken's bust of her; and a colossal wax ear and the inlaid table it sat on.[28]

The institute formally inaugurated the Manzolini Room in 1777. A great marble plaque marked the entrance to the Anatomy Museum, indicating in Latin that here one would find "The Celebrated Works of the Anatomy of the Human Body by Anna Morandi Manzolini." Significantly, administrators of the institute installed Morandi's waxworks in the first room, dedicated to anatomy, which immediately followed from the Physics Museum, or Room of Light—displacing to a second anatomy room Ercole Lelli's grandiose écorchés. The revised syntax of the expanded Anatomy Museum symbolized a striking reassessment of the cultural value of Morandi's and Lelli's rival anatomical models.

That Lelli's oeuvre had slipped from its high perch as the icon of Bolognese anatomical art was made explicit in Luigi Galvani's oration in June 1777 for the inauguration of the Manzolini Room.[29] The institute's newly named professor of anatomy and, from Lelli's death in 1766, the Custodian and Ostensor of the Anatomy Museum asserts the superiority of Morandi's comprehensive collection of anatomical models over Lelli's limited series of muscle men for teaching anatomy to medical students. For Galvani, it was both their didactic utility and their scientific promise that lent greater authority to Morandi's representations of the anatomized body. He contrasts the evident utility of her waxworks, which he himself employed to teach anatomy at the University of Bologna, with the near didactic irrelevance of Lelli's figures:

And so that you, my Listeners, comprehend the extent to which the Manzolini collection is useful for our youth, I beg you to transport your memory to past times when, because of the scarcity of anatomical tables [only Lelli's], we were forced to offer our students every year a sole course in Osteology and Myology. Remember how often they complained to us about having to always hear the same instruction on the bones and muscles.[30]

Morandi's precise, three-dimensional illustrations of the *living* body, from complete body systems to diaphanous, microscopic parts, were, in Galvani's estimation, devoid of those common impediments to sight and scientific gratification faced in the anatomy laboratory, and thus represented an unprecedented critical tool for students and, indeed, professors of anatomy. Her anatomical waxworks paradoxically freed the eye and the hand of the anatomist who employed them to know the body more completely than through dissection itself. In his eulogy, Galvani delights at once in Morandi's "realism," the experiential barriers she defied in her waxworks to make living anatomies and the affiliate extrascientific view of human anatomy she conceived: "these parts, true and natural, do not have, as cadavers do, any of the forbidding or putrid, which can nauseate or cause suffering in the tenderest of souls. On the contrary in their beauty and elegance they even stir fascination and award an almost incredible pleasure to those who study them."[31] Strongly reminiscent of Aristotle's notion of the "pleasure" and transformative power of artistic imitation to render "objects, which in themselves we view with pain, [objects of delight] when reproduced with minute fidelity,"[32] Galvani celebrates at once the authenticity and the radical newness of Morandi's waxworks.[33] In the professor's expert view, the waxes provided important and, indeed, unexpectedly delightful knowledge of the body by surpassing the work of conventional anatomists (like himself) in crucial ways.

> I declare that in these tables because of the way they are displayed and correlated, in a glance one can immediately comprehend and discern the natural form, the extension and the placement of the parts far better than what one can see when the parts are removed from cadavers and dissected in the way that for us Anatomists is usual. . . . Let us consider first off how much of the form and figure in those parts can remain true and natural, after tearing them from cadavers. Necessarily compressed by the hands, [the organs and parts] are drained of fluids and dry out at contact with the air. Their natural elasticity causes them to wrinkle and, by laying them on the table, they flatten and also become stretched out by the forceps and deformed when fastened by needles. . . . All of the parts of the cadaver prepared for use by Anatomists lose the placement, direction and extension of the natural figure. To all of this the tables that we are discussing give stability. If something is defective they restore it, such that you would not easily be able to establish if this art owes more to nature or nature to this art.[34]

Yet, despite his generous acknowledgment of Anna Morandi's remarkable gifts, her ability to overcome the limits of her sex to confront the dissection of fetid body parts and the complexities of anatomical science, as well as her sculpture of supremely accurate, useful, and indeed beautiful anatomical models, he does not include her finally among the brotherhood of anatomists, of which he was an elite member. Instead, he lauds her as an "extraordinary Artisan" (*Artefice straodinaria*).[35] She combined the talents of the painstaking *sector* and the assiduous and ingenious *ostensor*, but not those of the elite scholarly *lector*, custodian and interpreter of theoretical knowledge of the human body. Notwithstanding Luigi Galvani's implicit circumscription of Morandi's academic authority, in his tribute he unreservedly recognizes her eclectic genius in the practice of *new* anatomical methods and the *true* and pleasing representation of nature. The bodies she produced "drew to themselves the gaze of one and all" because they surmounted the limits of science and thereby reinvented the method, the means, and the aim of science itself.[36]

Postmortem

On 22 February 1775, a mere eight and a half months after Morandi's death and fifty miles from her native city, the Royal Museum of Physics and Natural History of Florence (the Specola) opened its doors to the public. Founded by Peter Leopold (1747–1792), Grand Duke of Tuscany, and directed by the natural philosopher Felice Fontana (1730–1805), the museum featured magnificently displayed collections of physics machines and chemistry instruments; innumerable native and exotic botanical, zoological, and mineralogical specimens; and, in six adjoining rooms, a vast exhibition of the most graphic, extensive, and spectacular anatomical wax models yet seen. The wax anatomies were a centerpiece of the museum and the result of some five years of intensive production by a large assemblage of anatomists, naturalists, artists, and sundry artisans and laborers. The latter included, among many others, sextons, who carted hundreds of cadavers for dissection and modeling across the city center from the Hospital of Santa Maria Nuova and the Orphanage of the Innocents. Carpenters, seamstresses, painters, iron- and goldsmiths, and glassmakers, who crafted the elegant display cabinets in which the waxworks were housed, also contributed their respective skills.[37]

Yet Fontana alone conceived and jealously micromanaged every aspect of the model production and exhibition. Daily logs that he ordered kept by his modelers, some of whom were only marginally literate, reveal in unvarnished terms the work conducted in the anatomy laboratories in Palazzo Torrigiani under his command. From these accounts and numerous other documents, it is clear that Fontana supervised the work of the hired dissector and strictly prescribed what parts of the body the modelers were to replicate, from what sources they were to copy these parts—whether actual bodies and/or canonical anatomical atlases—and how they were to be imaged. Frequent log entries refer to the corrections or complete refabrications Fontana ordered the modelers to undertake when the finished waxwork did not conform to his expectations. Less explicit though no less obvious in these daily chronicles is the modelers' deep animosity toward the tyrannical director who sought to control every particular of their work while relentlessly demeaning them.[38]

Their resentment was well placed. Fontana, as Anna Maerker has discussed, viewed the modelers—including those like the master wax sculptor Clemente Susini (1754–1814), who had received formal training in sculpture and design at Florence's prestigious Accademia delle Arti del Disegno—as ignorant, undisciplined manual workers, the grimy, toiling hired "hands" to his pristine theorizing mind in the realization of his scientific masterwork.[39] Fontana blatantly expresses his disdain for "my artists," as he called them, in a letter written to his colleague, the Bolognese anatomist and physiologist Leopoldo Marco Antonio Caldani. Lamenting the time and effort he had devoted to oversight of the production of his wax anatomy collection, he avers "my artists do not know how to write, to read, or to draw, and I use them as one would use a plane, a saw, or a scalpel."[40]

Caldani, the recipient of these remarks, had of course been an intermediary and protector of Anna Morandi's at the height of her career after her husband's death. It is, in fact, quite possible that Caldani had brought the young Fontana and the Lady Anatomist together at some point during Fontana's early years of science and anatomy study in Bologna from 1755 to 1758. The celebrated anatomical waxworks made by Morandi as well as those by Ercole Lelli were well known to Fontana, and a clear source of inspiration and the standard he aimed to surpass in his production and exhibition of the Florentine waxworks. As his letter to Caldani makes plain, however, he embraced an extremely hierarchical view of the relationship of science to art and of the natural philosopher to the artist, at direct odds with Morandi's art of anatomy. His concept of science and, specifically, anatomical science evokes a pre-Vesalian division of labor: the philosopher-lector, book in hand, presiding from his throne over those laboring to cut, probe, and image the putrefying corpse down below.[41]

With Morandi's death and the concurrent erection of Fontana's celebrated theater of knowledge, which immediately cast a shadow over Bologna's anatomy museum, the authority of the anatomical wax modeler came to an abrupt end. Yet, despite Fontana's radical dis-integration of the mental and manual labor, the work of the eye and hand, the theory and practice, the science and art that cohered seamlessly in Morandi's anatomical oeuvre and, perhaps more important, in her very person, the imprint of these alliances would remain discernible in the enduring bodies she formed in soft wax.

Notes

Introduction

1 Because she is identified at times by her maiden name, Morandi, and by her married name, Manzolini, in documents and secondary sources, I have included both. From this point onward, however, I will refer to her as Morandi, except in direct citations.

2 I use *studio* and *laboratory* interchangeably from this point, when discussing the space in which Manzolini and Morandi worked. It was, of course, a necessary and distinctive combination of science lab and art studio for the mixed work they conducted there.

3 The crucial original study of Bologna's tradition of scientific women in the early eighteenth century is *Alma mater studiorum: La presenza femminile dal XVIII a XX secolo; Ricerche sul rapporto Donna/Cultura Universitaria nell'Ateneo Bolognese* (Bologna: CLUEB, 1988). Although he never published his 1997 doctoral thesis, "La fortuna dei ceroplasti bolognesi del Settecento," Elio Vassena, MD, must be credited with producing the most comprehensive history to date of eighteenth-century Bolognese anatomical ceroplasty. I was not aware of this study until after I had already conducted extensive research on the history of anatomical wax modeling, but it proved invaluable for locating additional relevant archival documents. Foundational for my research was also the exhibition catalogue *Le cere anatomiche bolognesi del Settecento* (Bologna: CLUEB, 1981), particularly the innovative analysis of its editor, Maurizio Armaroli. The various contributions cited throughout this book on the rise of anatomical science and wax modeling were critical guideposts for my research. Of particular importance were the following: Franco Ruggeri, "Il Museo dell'Istituto di Anatomia Umana Normale," in *I luoghi del conoscere: I laboratori storici e i musei dell'Università di Bologna*, ed. Amilcare Pizzi (Bologna: Banca del Monte di Bologna e Ravenna, 1988); Raffaele A. Bernabeo, "Ercole Lelli" and "La libreria scientifica di Anna Morandi Manzolini," in *Le cere anatomiche bolognesi del Settecento*, ed. Maurizio Armaroli (Bologna: CLUEB, 1981), 30–32, 36–39; Mario

Fanti, "Sulla figura e l'opera di Marcello Oretti," *Il Carrobbio* 8 (1982): 125–43; Massimo Ferretti, "Il notomista e il canonico," in *I materiali dell'Istituto delle Scienze* (Bologna: CLUEB, 1979), 100–114; Walter Tega, "Mens agitat molem: L'Accademia delle Scienze di Bologna (1711–1804)," in *Scienza e letteratura nella cultura italiana del Settecento*, ed. Renzo Cremente and Walter Tega (Bologna: Il Mulino, 1984), 65–108; Tega, ed., *I commentari dell'Accademia delle Scienze di Bologna*, vol. 1 of *Anatomie accademiche* (Bologna: Il Mulino, 1986); and Annarita Angelini, ed., *L'Istituto delle Scienze di Bologna*, vol. 3 of *Anatomie accademiche* (Bologna: Il Mulino, 1993). My analysis of Anna Morandi's role and influence within the eighteenth-century Bolognese cultural milieu also necessarily drew on the pathbreaking work of Marta Cavazza on the Bolognese history of science as well as the roles played by women within that history. I found especially useful her book *Settecento inquieto: Alle origini dell'Istituto delle Scienze di Bologna* (Bologna: Il Mulino, 1990), as well as a number of critical articles: "Dottrici e lettrici dell'Università di Bologna nel Settecento," *Annali di storia delle università italiane* 1 (1997): 109–26; "Women's Dialectics, or the Thinking Uterus: An Eighteenth-Century Controversy on Gender and Education," in *The Faces of Nature in Enlightenment Europe*, ed. Lorraine Daston and Gianna Pomata (Berlin: BWV, 2003), 237–57; and "La recezione della teoria halleriana dell'irritabilità nell'Accademia delle Scienze di Bologna," *Nuncius: Journal of the History of Science* 12, no. 2 (1997): 359–77. Of course, Paula Findlen's studies of the rise of learned women in Settecento Bologna were also crucial for the development of my vision of Morandi at work in this city: "Science as a Career in Enlightenment Italy: The Strategies of Laura Bassi," *Isis* 84, no. 3 (1993): 441–69; and "Translating the New Science: Women and the Circulation of Knowledge in Enlightenment Italy," *Configurations* 3, no. 2 (1995): 167–206. Findlen's pathbreaking work on the early history of collecting and the development of natural history museums in Bologna proved pivotal for my analysis of the rise of anatomical wax museums: *Possessing Nature: Museums, Collecting and Scientific Culture in Early Modern Italy* (Berkeley and Los Angeles: University of California Press, 1994).

4 Biblioteca comunale dell'Archiginnasio Bologna (hereafter BCAB), MS B 133, vol. 11, Marcello Oretti, *Notizie de professori del disegno cioè pittori, scultori ed architetti bolognesi e de' forestieri di sua scuola raccolte ed in più tomi divise da Marcello Oretti bolognese accademico dell'Istituto delle Scienze di Bologna*, fol. 229.

5 Morandi had an extensive personal archive of Latin texts on medicine and anatomy, including works by Casseri, Cowper, Valverde, Morgagni, Valsalva, Vesalius, Bartholini, and Malpighi. Her many explicit references to these texts in her anatomical notebook make it clear that she had deep knowledge of their contents. The list of works in her library is found in a separate section after the close of her anatomical notebook: Biblioteca Universitaria, Bologna (hereafter BUB), MS 2193, *Catalogo delle preparazioni anatomiche in cera formanti il gabinetto anatomico prima della Reggia Università* (henceforth cited as Morandi, *Anatomical Notebook*).

6 Ibid., fol. 227.

7 Archivio Arcivescovile di Bologna (hereafter AAB), *Parocchie di Bologna soppresse*, San Niccolò degli Albari, Cart. 35/8, 3. Cited by Gabriella Berti Logan, "Women and the Practice and Teaching of Medicine in Bologna in the Eighteenth and Early Nineteenth Centuries," *Bulletin of the History of Medicine* 77, no. 3 (2003): 512n12.

8 On this, see Cavazza, *Settecento inquieto*, 40–41.

9 Amedeo Benati, Antonio Ferri, and Giancarlo Roversi, *Storia di Bologna* (Bologna: ALFA, 1978), 251.

10 BUB, MS 75, II, cc. 323–38. Anton Felice Marsili, *Discorso dell'apertura delle due accademie in casa di Mons. Arcidiacono Ant. Felice Marsili, l'una ecclesiastica e l'altra filosofica nel mese di novemb. 1687 e dal medesimo Monsignore derivato.* Cited by Cavazza, *Settecento inquieto*, 92.

11 Luigi Ferdinando Marsili, *A tutti gli ordini della Città di Bologna* (1711). Quoted in Angelini, *L'Istituto delle Scienze*, 3:98.

12 Cavazza, *Settecento inquieto*, 31.

13 Anton W. A. Boschloo, *L'Accademia Clementina e la preoccupazione del passato* (Bologna: Nuova Alfa Editoriale, 1989), 14.

14 Angelini, *L'Istituto delle Scienze*, 3:135.

15 Ibid., 3:115–39.

16 Lambertini was custodian of the Vatican Library from 1712 to 1726. On this dispute see Findlen, "Science as a Career in Enlightenment Italy," 458; and John Stoye, *Marsigli's Europe, 1680–1730: The Life and Times of Luigi Ferdinando Marsigli, Soldier and Virtuoso* (New Haven, CT: Yale University Press, 1994), 304–9.

17 There are no adequate studies of the life and influence of Pope Benedict XIV. A dated biographical overview can be found in Renée Haynes, *The Philosopher King: The Humanist Pope Benedict XIV* (London: Weidenfeld and Nicolson, 1970). Christopher Johns is completing a book titled *The Visual Culture of Catholic Enlightenment* that is sure to be a crucial contribution to studies of Benedict, especially his patronage of the arts.

18 "Se Iddio ci aiuterà, che prima di morire possiamo dargli un poco di dote, l'Istituto è in grado di poter rendere celebre la nostra patria, come in altri tempi fu celebre per l'università." Prospero Lambertini to Paolo Magnani, letter of 2 September 1744, cited in Tega, *I commentari dell'Accademia delle Scienze di Bologna*, 1:31.

19 Tega, "Introduzione," ibid., 1:34–35.

20 *Notificazione de cadaverum sectione facienda in publicis Academiis, utrum constitution Bonifacii VIII adversetur.* This document, which issued from the palace of the archbishop in the vernacular, was republished numerous times over the century and was translated into Latin and Spanish. See G. Martinotti, *Prospero Lambertini (Benedetto XIV) e lo studio dell'anatomia in Bologna* (Bologna: Tipografica Azzoguidi, 1911), 4–5.

21 For in-depth and challenging analyses of the controversial papal bull, see the following: Elizabeth A. Brown, "Authority, the Family and the Dead in Late Medieval France," *French Historical Studies*, 16, no. 4 (Fall 1990): 803–32; and Agostino Paravicini Bagliani, "Storia della scienza e storia della mentalità: Ruggero Bacone, Bonifacio VIII e la teologia della prolugatio vitai," in *Aspetti della letteratura latina nel secolo XIII: Atti del primo convegno internazionale di studi dell'Associazione per il medioevo e l'umanesimo latini (AMUL)* 15 (1986): 243–80.

22 "Con questa Nostra Notificazione facciamo sapere, che, quando non si tratta de' cadaveri di giustiziati [...] ma de' cadaveri d'uomini, o di donne di qualsivoglia condizione, morti di qualunque altra morte, che si credano necessari per la Notomia da farsi nelle pubbliche Scuole, se ne faccia o a Noi, o al Nostro Vicario Generale l'istanza, con sicurezza, che, per non impedire un'opera tanto utile, si

prenderanno tutte le misure opportune e per il consenso de' parenti, e per il diritto del Parroco, e per l'Esequie." Martinotti, *Prospero Lambertini*, 4–5.

23 Claudia Pancino, "L'ostetrica del Settecento e la scuola bolognese di Giovanni Antonio Galli," in *Ars obstetricia bononiensis: Catalogo ed inventario del Museo Obstetrico Giovan Antonio Galli* (Bologna: CLUEB, 1988), 28.

24 In Oretti's words, "[Anna Morandi] made for Dr. Galli a great series of uteruses and fetuses admirably formed that later passed to a room on the bottom floor of the Institute and that continue to serve a school of scholars of Medicine who exercise their talent in this great Mother of Learning." BCAB, MS B 133, vol. 11, fol. 227.

25 BCAB, MS B 134, vol. 12, Oretti, *Notizie de' professori dell'arte del disegno*, "Giovanni Manzolini," fol. 135; Luigi Crespi, *Felsina pittrice: Vite de' pittori bolognesi* (Rome: Stamperia di Marco Pagliarini, 1769; reprint, Bologna: Arnaldo Forni, 1980), 306–7. The pope established the courses in Italy on "the demonstration of surgical operations on cadavers," with Molinelli as the head on 23 August 1742.

26 Olschki has just published the first modern edition of Morandi's notebook—Miriam Focaccia, ed., *Anna Morandi Manzolini: Una donna fra arte e scienza; Immagini, documenti, repertorio anatomico* (Florence: Olschki, 2008).

27 Barbara Maria Stafford, *Body Criticism: Imaging the Unseen in Enlightenment Art and Medicine* (Cambridge, MA: MIT Press, 1991), 53.

28 Archivio di Stato, Bologna (hereafter ASB), Archivio Ranuzzi, *Istrumenti scritture diverse spettanti alla Nobile Casa Ranuzzi dall'anno 1769 all'anno 1773*, Libro 124, n. 21, "Inventario dei libri venduti dalla Signora Manzolini al Sig.r Sen.re Girolamo Ranuzzi, con il suo Armario."

29 See especially Findlen, "Science as a Career in Enlightenment Italy," 441–69; Findlen, "Translating the New Science," 167–206; Marta Cavazza, "Laura Bassi," Bologna Science Classics On-line: http://www.137.204.24.205/cis13b/bsc03/bassi/bassinotbyed/bassinotbyed.pdf; and Cavazza, "Dottrici e lettrici dell'Università di Bologna nel Settecento," *Annali di storia delle università italiane* 1 (1997): 109–26.

30 Thomas Laqueur, *Making Sex: Body and Gender from the Greeks to Freud* (Cambridge, MA: Harvard University Press, 1990). The two sides of the debate over Laqueur's theories on the emergence during the eighteenth century of sexually differentiated anatomy are succinctly presented in volume 94 of *Isis* (2003). See in this volume Michael Stolberg, "A Woman Down to Her Bones: The Anatomy of Sexual Difference in the Sixteenth and Early Seventeenth Centuries," 274–99; Thomas Laqueur, "Sex in the Flesh," 300–306; and Londa Schiebinger, "Skelettestreit," 307–13.

31 I use this term guardedly. While some of the modelers were formally trained artists and gained membership to the prestigious Academy of Fine Art in Florence, all were viewed by Fontana as artisans.

32 Anna Maerker has contributed important analyses of the real and symbolic relationship between the natural philosopher Fontana and the artists and artisans in his command. See "Model Experts: The Production of Anatomical Models at La Specola, Florence and the Josephinum, Vienna, 1775–1814," Ph.D. diss., Cornell University, 2005; and "The Anatomical Models of the Specola: Production, Uses, and Reception," *Nuncius: Journal of the History of Science* 21, no. 2 (2006): 295–321.

33 Lucia Dacome builds on Pamela's Smith's interpretation of the alchemical associa-

tions in artisanal knowledge and practices (Pamela H. Smith, *The Body of the Artisan* [Chicago: University of Chicago Press, 2004]). See Lucia Dacome, "'Un certo e quasi incredibile piacere': Cera e Anatomia nel Settecento," *Intersezioni: Rivista di storia delle idee* 25 (December 2005): 415–36; "Waxworks and the Performance of Anatomy in Mid-18th-century Italy," *Endeavor* 30, no. 1 (March 2006): 29–35; and "Women, Wax and Anatomy," *Renaissance Studies* 21, no. 4 (September 2007): 522–50.

34 Focaccia, introduction, "Anna Morandi Manzolini: Una Donna Fra Arte E Scienza," in *Anna Morandi Manzolini*.

35 Maerker, "The Anatomical Models of the Specola; and "Uses and Publics of the Anatomical Model Collections of La Specola, Florence, and the Josephinum, Vienna, around 1800," in *From Private to Public. Natural Collections and Museums*, ed. Marco Beretta (Nantucket, MA: Science History Publications, 2005), 81–96; Simone Contardi, *La casa di Salomone a Firenze: L'Imperiale e Reale Museo di fisica e storia naturale 1775–1801* (Florence: Olschki, 2002).

36 Petronio Ignazio Zecchini, *Dì geniali: Della dialettica delle donne ridotta al suo vero principio* (Bologna: A. S. Tommaso d'Aquino, 1771).

37 Giovanni Bianchi, "Letter Written from Bologna to a Friend in Florence 21 September 1754," in *Novelle letterarie di Firenze*, 15:708–11 (Florence: Marco Lastri, 1740–92).

Chapter 1

1 On Lambertini's tenure as Bologna's archbishop, see Mario Fanti, "Prospero Lambertini arcivescovo di Bologna (1731–1740)," in *Benedetto XIV* (Cento: Centro Studi Girolamo Baruffaldi, 1982), 165–233.

2 Walter Tega has written eloquently on the realliance under Pope Lambertini of science and faith in the introduction of *I commentari dell'Accademia delle Scienze di Bologna* (Bologna: Il Mulino, 1986), 27–32, as well as in "Papa Lambertini: Una lucida visione dei rapporti fede-scienza," *Secularia Nona* 13 (1996–97): 92–98.

3 My thanks to the students and faculty in the History of Science and Museums program at Johns Hopkins University, especially Elizabeth Rodini and Gianna Pomata, for their insights on the various connotations of the space of the Anatomy Museum.

4 This is the tripartite doctrine of scientific practice set down by Luigi Ferdinando Marsili for his Institute of Science.

5 On the history and cultural and symbolic significance of wax modeling in art and science, see Julius Ritter von Schlosser, Thomas Medicus, Edouard Pommier, and Gotthold Ephraim Lessing, *Histoire du portrait en cire* (Paris: Macula, 1997); Julius Ritter von Schlosser, *Geschichte der Porträtbildnerei in Wachs* (Leipzig: Wien, 1911); Mario Paz, "Le figure di cera in letteratura," in *La ceroplastica nella scienza e nell'arte: Atti del Primo Congresso Internazionale* (Florence: Olschki, 1975), 549–68; Ernst H. Gombrich, *Art and Illusion: A Study in the Psychology of Pictorial Representation* (Princeton, NJ: Princeton University Press, 1969); and Roberta Panzanelli, *Ephemeral Bodies: Wax Sculpture and the Human Figure* (Los Angeles: Getty Research Center, 2008). On wax modeling in Bologna, see Alessandro Ruggeri and Anna

Maria Bertoli Barsotti, "The Birth of Waxwork Modelling in Bologna," *Italian Journal of Anatomy and Embryology* 102, no. 2 (April–June 1997): 99–107.

6 On the notion of science as stage-managed public drama to bolster the authority of particular interest groups within the academic or the political sphere, see Stephen Hilgartner, *Science on Stage: Expert Advice as Public Drama* (Stanford, CA: Stanford University Press, 2000), 3–23.

7 Giovanna Ferrari, "Public Anatomy Lessons and the Carnival: The Anatomy Theatre of Bologna," *Past and Present* 117 (November 1987): 52, 96. Ferrari provides a detailed overview of how the public anatomy lesson developed from a didactic demonstration by university professors for their students into a public ceremony and cultural spectacle.

8 Dacome asserts a connection between the Carnival and wax anatomy; however, she places a very different emphasis on the transgressive nature of both, and Anna Morandi's carnivalesque upturning of roles in her anatomy studio: "'Un certo e quasi incredibile piacere,'" 415–36.

9 Lorraine Daston and Peter Galison, *Objectivity* (New York: Zone Books, 2007), 42.

10 Findlen, "Science as a Career in Enlightenment Italy," 441–69; Findlen, "Translating the New Science," 167–206; Cavazza, "Laura Bassi"; and Cavazza, "Dottrici e lettrici," 109–26.

11 While Dacome highlights the broad correspondence and continuity between the production of wax models and the performance of anatomical spectacle by Ercole Lelli and Anna Morandi (working together with her husband, Giovanni Manzolini) within the context of Lambertini's cultural revival of Bologna, in this chapter and through the course of the book, my focus centers instead on the discontinuity between Anna Morandi's work in anatomical science outside the cultural mainstream and the art of anatomy epitomized by the pope's chosen modeler, Ercole Lelli. I contend that the difference that Morandi presents to authorized visions of the body, including the body of the anatomist, both defines her lifework and elucidates critical traits and tensions of the dominant cultural milieu. See Dacome, "'Un certo e quasi incredibile piacere,'" 415–35.

12 On the subject of the use of criminal bodies for the Public Anatomy, see Ferrari, "Public Anatomy Lessons and the Carnival," 58–69; and Andrea Carlino, *Books of the Body: Anatomical Ritual and Renaissance Learning*, trans. John Tedeschi and Anne C. Tedeschi (Chicago: University of Chicago Press, 1999), 92–108. Carlino indicates that the anatomy could be held as much as five to ten days after the execution. Adriano Prosperi has recently unearthed documents that indicate an often strategic effort on the part of eighteenth-century Italian civic officials to set or even delay the execution date in order for it to take place the very day of the Public Anatomy. My thanks to Professor Prosperi for communicating his findings to me.

13 Lionello Puppi, *Torment in Art: Pain, Violence and Martyrdom* (New York: Rizzoli, 1991), 51. On the cultural history and symbolism of torture and execution, see also the following by Adriano Prosperi: *Dare l'anima: Storia di un infanticidio* (Turin: Einaudi, 2005); "Il sangue e l'anima: Ricerche sulle Compagnie di Giustizia in Italia," *Quaderni Storici* 51, no. 17 (1982): 959–99; and "Esecuzioni captiali e controllo sociale nella prima età moderna," in *La pena di morte nel mondo: Convegno Internazionale Bologna* (Casale Monferrato, Italy: Marietti, 1983), as well as Edward Muir, *Civic Ritual in Renaissance Venice* (Princeton, NJ: Princeton University Press,

1981); William F. Schulz, ed., *The Phenomenon of Torture: Readings and Commentary* (Philadelphia: University of Pennsylvania Press, 2007); David Freedberg, "The Representation of Martyrdoms during the Counter Reformation in Antwerp," *Burlington Magazine* 118 (1976): 128–38; Pieter Spierenburg, *The Spectacle of Suffering: Executions and the Evolution of Repression from a Preindustrial Metropolis to the European Experience* (Cambridge: Cambridge University Press, 1984); Michel Foucault, *Discipline and Punish: The Birth of the Prison*, trans. Alan Sheridan (New York: Vintage Books, 1977); and Michael Sappol, *A Traffic of Dead Bodies* (Princeton, NJ: Princeton University Press, 2002).

14 Adriano Baccilieri, "La storia dell'edificio: Dall'Ospedale di S. Maria della Morte al Museo Civico," in *Dalla Stanza della Antichità al Museo Civico*, ed. Cristiana Morigi Govi and Giuseppe Sassatelli (Bologna: Grafis Edizioni, 1984), 101–14.

15 ASB, Assunteria di Studio, *Diversorum*, busta 91, "Cerimoniale per le lezioni di anatomia," 1765.

16 My account of this ritual event is deeply indebted to Ferrari's original and masterly description of it. Ferrari, "Public Anatomy Lessons and the Carnival," 50–51.

17 On Bassi's role as civic spectacle, see the recent article by Marta Cavazza, "Between Modesty and Spectacle: Women and Science in Eighteenth-Century Italy," in *Italy's Eighteenth Century: Gender and Culture in the Age of the Grand Tour*, ed. Paula Findlen, Wendy Wassing Roworth, and Catherine Sama (Stanford, CA: Stanford University Press, 2009), 275–302. See also Findlen, "Science as a Career in Enlightenment Italy," 451.

18 See especially Ferrari, "Public Anatomy Lessons and the Carnival," 50–106; Carlino, *Books of the Body*; William S. Heckscher, *Rembrandt's Anatomy of Dr. Nicolaas Tulp: An Iconographical Study* (New York: New York University Press, 1958); and Andrew Cunningham, "The End of the Sacred Ritual of Anatomy," *Canadian Bulletin of Medical History* 18 (2001): 187–204.

19 Cunningham, "The End of Sacred Ritual of Anatomy," 193.

20 Cited in the translated form here by Giovanna Ferrari in her article "Public Anatomy Lessons and the Carnival," 75. The citation was taken from ASB, Gabella Grossa, *Libri Segreti*, 9 October 1637, fol. 237.

21 ASB, Assunteria di Studio, *Diversorum*, busta 91, "Anatomia Pubblica," May 1755.

22 Heckscher, *Rembrandt's Anatomy of Dr. Nicolaas Tulp*, 27–34.

23 See Carlino, *Books of the Body*, 109–19; Puppi, *Torment in Art*; Prosperi, *Dare l'anima*; and Prosperi, "Esecuzioni capitali."

24 Prosperi, *Dare l'anima*, 316.

25 On this subject see also Katharine Park, *Secrets of Women: Gender, Generation, and the Origins of Human Dissection* (New York: Zone Books, 2006), esp. 221–34.

26 Ibid., 24.

27 It is important to note here that this was the only "monstrous" body represented in Lelli's oeuvre and was assigned to him by the Academy of Science in the Institute. All other images of the body by Lelli or, for that matter, by Anna Morandi and Giovanni Manzolini were strictly "normal" universal types.

28 ASB, Assunteria d'Istituto, *Atti*, vol. 3 (1727–34). Lelli did not follow this directive exactly, but frequently built his wax models on actual human bones.

29 ASB, Assunteria d'Istituto, *Diversorum*, busta 10, "Proposta per le statue anatomiche," 12 February 1732:

The statues exhibited to represent the entire musculature of the human body in its natural state will number five. The first will be of the deepest muscles attached and adjacent to the bone and, in order to better represent these, the bones of an actual skeleton will form the base of the statue. In this [first statue] the scalene muscles, the internal and external intercostals, the supra-costals . . . the psoas muscles, the internal iliac muscles, and many others will be demonstrated . . . . The second statue will represent the muscles overlaid on those already described, and to this statue will also be added the lower and upper limbs with the abdomen covered by the peritoneum in which the transversus and rectus muscles are seen together with other muscles of the torso . . . . The third statue will express the muscles overlaid on those of the second statue and for this reason in this statue will be seen within the abdomen the ascending obliquus muscles, the pyramidalus, the serratus anterior and posterior, the quadratos of the neck, and the complexus [semispinalis], the rectus and obliquus of the head . . . . The fourth statue will demonstrate in the torso the descending oblique muscles, the pectoralis minor, the rhomboideus, the splenius, and that of . . . the levator scapulae, the coracobrachialis, ulnar and radial muscles in the upper limbs, in the lower, the solar, the plantar, the perforantes and the extensor brevis. The fifth will demonstrate in the necessary order the complete external musculature . . . . Thus, with the five statues together we will have a complete representation of the muscular system.

30 Born in 1702 to Monica Tagliaferri and Domenico Maria Lelli, a craftsman of portable firearms, Lelli became famous at an early age for his exquisite relief engraving of the heavy matchlock guns. Although he clearly preferred the art of engraving to the mechanics of gun making, his artistic ambitions greatly exceeded those that could be realized in his father's bottega. He sought formal training on his own in the fine arts, and shrewdly managed to apprentice himself to leading local masters of painting, architecture, ornamental design, and sculpture.

31 Bernabeo, "Ercole Lelli," 30.

32 Ibid.

33 Giampietro Zanotti, "Al Sig. Ercole Lelli da Giampietro Zanotti, in Giovanni Gaetano Bottari," in *Raccolta di lettere sulla pittura, scultura ed architettura scritte da' più celebri personaggi dei secoli XV, XVI, e XVII, pubblicata da M. Gio. Bottari, e continuata fino ai nostri giorni da Stefano Ticozzi* [1757] (Milan: G. Silvestri, 1822–25), 2:193.

34 Ibid., 2:64. Notwithstanding the circumspect view of anatomical modeling taken at the conservative Clementina, artists in the academy rushed to copy the miniature, and numerous plaster casts were made of it and distributed throughout the city. See Michele Medici, "Elogio d'Ercole Lelli," in *Memorie della Accademia delle Scienze dell'Istituto di Bologna* (Bologna: San Tommaso d'Aquino, 1856), 7: 158–86; and Zofia Ameisenowa, *The Problem of the Écorchés and the Three Anatomical Models in the Jagiellonian Library,* trans. Andrezej Potocki (Warsaw: Zaklad Narodowy Imtenja Ossolinskich, 1963), 63–64.

35 As Giovanna Ferrari has observed, the anatomical theater in the Archiginnasio was viewed by the Senate and university administrators as "extremely symbolic, a kind of sum total of the glories of the Studium," which would exert a new and vigorous pull on scholars, foreign and native, and thereby reverse the declining

stature of the city and the university: "Public Anatomy Lessons and the Carnival," 81.

36 Giovanni Fantuzzi, *Notizie degli scrittori bolognesi* (Bologna: Stamperia di Tommaso d'Aquino, 1786), 5:50–51.

37 There is no adequate term in English to describe the *attaccato campanilismo* to which Lambertini was given. It literally means an avid devotion to the parish bell tower.

38 Benedict XIV, *De servorum dei beatificatione et beatorum canonizatione* (Bologna: Longhi, 1734), 4:384.

39 Marta Cavazza's detailed explanation of this and other related insignia proved invaluable for my own interpretation. See her article "Laura Bassi."

40 The families represented are the Banzi, Gessi Rossi, Gioannetti, Sampieri, Orsi, Marsili-Rossi, Grassi, and Vassé-Pietramellara. Giuseppe Plessi, *Le Insignia degli Anziani del Comune dal 1530–1796 Catalogo-Inventario* (Rome: Archivio di Stato di Bologna, 1954), 245.

41 Minerva was the traditional symbolic figure of the city to which Bologna's leaders explicitly connected Bassi.

42 Findlen, "Science as a Career in Enlightenment Italy," 448.

43 Plessi, *Le Insignia degli Anziani del Comune*, 245. Marta Cavazza, "Laura Bassi," Bologna Science Classics Online: www.cis.unibo.it/cis13b/bsco3/bassi/bassinotbyed/bassinotbyed.pdf.

44 Cavazza, "Dottrici e lettrici," 115.

45 Most frequent among these female exemplars were Bitisia Gozzadini (1209–1261), reputedly degreed in law in 1237 and who, dressed always in men's robes, taught jurisprudence for years at the university; the sisters Novella and Bettina Calderini, who, in the fourteenth century, also taught philosophy and law; the anatomist Mondino de Liuzzi's assistant in anatomical dissection, Alessandra Giliani (d. 1326); the famously reticent Novella d'Andrea, who taught hidden behind a curtain for her father, the cardinal and monastic reformer Giovanni d'Andrea, and whose mother, Milanzia dell'Ospedale, held a chair in legal studies; and Dorotea Bocchi, degreed in medicine in 1436, who took over her father's university teaching after his death.

46 On Roccati, see Paula Findlen, "Translating the New Science," 167–206, and "Becoming a Scientist: Gender and Knowledge in Eighteenth-Century Italy," in "Scientific Personae," ed. Lorraine Daston and Otto Sebum; special issue, *Science in Context* 16 (2003): 59–87. See also Maria Laura Soppelsa and Eva Viani, "Dal newtonianismo per le dame al newtonianismo delle dame: Cristina Roccati una 'savante' del Settecento Veneto," in *Donne, filosofia e cultura nel Seicento*, ed. Pina Totaro (Rome: Consiglio Nazionale delle Ricerche, 1999), 211–40.

47 On Agnesi, see Massimo Mazzotti, "Maria Gaetana Angesi: Mathematics and the Making of the Catholic Enlightenment," *Isis* 92 (2001): 657–83. See also Paula Findlen's biographical overview that introduces her translation of Agnesi's oration in defense of women's learning, "Maria Gaetana Agnesi, Translator's Introduction," in *The Contest for Knowledge*, ed. Rebecca Messbarger and Paula Findlen (Chicago: University of Chicago Press, 2005), 117–27.

48 Cavazza, "Dottrici e lettrici," 115.

49 See Rebecca Messbarger, *The Century of Women: Representations of Women in Eighteenth-Century Italian Public Discourse* (Toronto: University of Toronto Press, 2002).

50 On Voltaire's admiration for Bassi, see Ernesto Masi, "Laura Bassi ed il Voltaire," in *Studi e ritratti* (Bologna: Zanichelli, 1881), 157–71. See also Joseph-Jérôme de Lalande, *Voyage d'un François en Italie fait dans les années 1765 & 1766* (Venice, 1767), 2:117; Johann J. Volkmann, *Historisch-kritische Nachtrichten von Italien, welche eine Beschreibung dieses Landes. Der Sitten, Reigierungsform, Handlung, des Zustandes des Wissenschaften und insonderheit der Kunst enthalten*, 3 vols. (Leipzig: Caspar Fritsch, 1777); and John Morgan, *Journal of Dr. John Morgan of Philadelphia from the City of Rome to the City of London 1764* (Philadelphia: J. B. Lippincott, 1907).

51 Charles Burney, *An Eighteenth-Century Musical Tour in France and Italy*, ed. Percy A. Scholes (London: Oxford University Press, 1959), 159.

52 Harvey Cushing, "Ercole Lelli and his Écorché," *Yale Journal of Biology and Medicine* 9, no. 3 (1937): 208.

53 On the symbolism of the Medusa referenced here, see Jonathon Sawday, *The Body Emblazoned: Dissection and the Human Body in Renaissance Culture* (New York: Routledge, 1995), 6–15. See also Marjorie Garber and Nancy J. Vickers, *The Medusa Reader* (New York: Routledge, 2003).

54 ASB, Fondo Senato, Serie Partiti, anni 1731–37, vol. 49, fol. 49v; cited in Elio Melli, "Laura Bassi Verati: Ridiscussioni e nuovi spunti," in *Alma mater studiorum*, 74.

55 Among the most noteworthy to study with her were her cousin Lazzaro Spallanzani, the future Chair of Natural History at Pavia and author of the important *Dissertazioni di fisica animale e vegetale* (Modena, 1780); the naturalist and physiologist Felice Fontana, who became court physician to Grand-Duke Peter Leopold and is perhaps best remembered for his graphic wax anatomical models made for the Florentine Specola; and the anatomist Leopoldo Marc Antonio Caldani, who, along with Fontana, was a leading defender of Haller's theories of irritability, theories tested on the electrical machine at Bassi's studio. See especially Cavazza, "Laura Bassi 'Maestra' di Spallanzani," in *Il cerchio della vita: Materiali di ricerca del Centro Studi Lazzaro Spallazani di Scandiano sulla storia della scienza del Settecento*, ed. Walter Barnardi and Paola Manzini (Florence: Leo S. Olschki, 1999), 185–202; and Cavazza, "Lesbia e Laura: Donne spettatrici e donne sperimentatrici nell'Italia del Settecento," in *Lorenzo Mascheroni: Scienza e letteratura nell'età dei lumi* (Bergamo: Bergamo University Press, 2004), 162–63.

56 The commission had obviously expanded from the five figures first proposed ten years earlier. The contract states that the pope,

> at the impulse of his generous heart toward the fine arts and especially toward the Institute of Sciences in this his city of Bologna . . . decided to erect at his own expense in this Institute an anatomical room in which would be shown, in figures both whole and partial, and distinctly made in wax and colored according to nature, the entire fabric of the human body, including the viscera that is exclusively called myology and osteology.

> ASB, Assunteria di Istituto, *Diversorum*, busta 10, no. 2, "Fondazione delle Camere di Anatomia fatte da Benedetto XIV 1742–1747 statue anatomiche formate in cera da Ercole Lelli," 1 December 1742.

57 Ibid.

58 Ibid.

59   Dacome addresses the subject briefly in "'Un certo e quasi incredibile piacere,'" 422–23.

60   See Messbarger, "Waxing Poetic" and "Re-membering a Body of Work," as well as the introduction to this book and Ludmilla Jordanova's article "Sex and Gender" in Christopher Fox, Roy Porter, and Robert Wokler, eds., *Inventing Human Science: Eighteenth-Century Domains* (Berkeley and Los Angeles: University of California Press, 1995), 14–19, 152–83.

61   On this subject, see Rebecca Messbarger, *The Century of Women: Representations of Women in Eighteenth-Century Italian Public Discourse* (Toronto: University of Toronto Press, 2002).

62   Londa Schiebinger, "Skeletons in the Closet: The First Illustrations of the Female Skeleton in Eighteenth-Century Anatomy," in *The Making of the Modern Body*, ed. Catherine Gallagher and Thomas Laqueur (Berkeley and Los Angeles: University of California Press, 1987), 42–82.

63   ASB, Assunteria d'Istituto, *Diversorum*, busta 10, no. 1, "Ercole Lelli, All'illustrissima ed eccellentissma Assunteria dell'Istituto delle Scienze di Bologna—Proposta per la nuova Camera Anatomica," 5 October 1742.

64   ASB, Assunteria dei Magistrati, *Affari Diversi*, busta 78, no. 3, "Benedictus Papa XIV Motu Proprio," 28 November 1747.

65   Ibid.

66   ASB, Assunteria d'Istituto, *Diversorum*, busta 10, no. 1, "Proposta per le statue anatomiche," 12 February 1732.

67   In his *Anatomical Compendium for Use in the Art of Design* that Lelli wrote and published late in life, he advises artists to emulate those "most famous" Greek sculptors, Agasias of the "stupendous Fighting Gladiator" and the three Rhodesian sculptors of the "famossissimo Laocoonte," whose knowledge of anatomy allowed them to create structurally perfect nudes roiling with life and passion. BCAB, MS B 1562, fols. 9–10, Ercole Lelli, *Compendio anatomico per uso delle Arti del disegno di Ercole Lelli bolognese pittore, scultore e Disegnatore famosissimo.*

68   William Cowper, "Hope," line 58, in *The Poetical Works of William Cowper*, ed. Nicholas Harris Nicolas (Edinburgh: William P. Nimmo, 1863), 174.

69   For a critical analysis of questions of gender in Albinus's study of skeletal structure, see Schiebinger, "Skeletons in the Closet."

70   See letter from Benedict XIV to Fr. Peggi, Rome, 6 May 1752, in *Briefe Benedicts XIV an den Canonicus Pier Francesco Peggi in Bologna (1729–1758)*, ed. Franza Xaver Kraus (Freiburg: J. C. B. Mohr, 1888), 89.

71   Lelli was awarded 560 *lire* and reimbursement for his expenses—from the Congregation of the Gabella Grossa, the bureaucratic body in charge of the financial management of the university; the Senate awarded him the post of Coiner of the Public Mint; poems circulated in his honor; and he attained an unprecedented position at the juncture of art and science in the *new* Bologna. On this see Cushing, "Ercole Lelli and his Écorché."

72   Boschloo, *L'Accademia Clementina*, 14.

73   Massimo Ferretti, "Il notomista e il canonico," 102.

74   Crespi, *Felsina pittrice*, 159–60. It should be noted that the diatribe appeared in print in 1769, several years after Lelli's death, but nonetheless reflects a dominant

strain of opinion during the ongoing controversy about Lelli and the specialization by Bolognese artists in anatomical science.

75 Carlo Pisarri, *Dialoghi tra Claro e Sarpiri per istruire chi desidera d'essere un eccellente
    pittore figurista* (Bologna: F. Pisarri, 1778); Gian Lodovico Bianconi, *Lettere del consi-
    gliere Gian Lodovico Bianconi scritte in nome del Segretario di S. Luca di Roma al segretario
    dell'Accademia Clementina di Bologna sopra il libro del canonico Luigi Crespi bolognese
    intitolato Felsina pittrice . . .* (Milan: Tipografia de' Classici Italiani, 1802).

76 Francesco Algarotti, *An Essay on Painting Written in Italian by Count Algarotti*
    (London: L. Davis and C. Reymers, 1763), 15.

77 Ibid., 15–16; this is a paraphrase.

78 Ibid., 16.

79 Eugenio Riccomini, *Mostra della scultura bolognese del Settecento* (Bologna: Tamari,
    1966), 104–8.

80 Legitimate questions loomed throughout Lelli's career regarding the extent to
    which he relied on assistants for the execution of major projects, beginning
    with his celebrated wooden *écorchés* in the Archiginnasio Anatomy Theater.
    The papal commission, unlike the Aldovrandi project, authorized and funded
    Lelli's employment of a sculptor to help with the wax modeling and a surgeon
    to retrieve body parts and conduct the anatomical dissections. When Lelli began
    work on the museum commission at the start of 1743, he hired the young sculptor
    Filippo Scandellari (1717–1802) to assist him, as stipulated in the original contracts.
    Scandellari remained with Lelli for only a short time, quickly replaced by the
    sculptor Angelo Piò's twenty-three-year-old son, Domenico (1720–1799). He too
    remained with Lelli only briefly, followed after little more than a year by Giovanni
    Manzolini. Documents shed little light on why these young artists promptly quit
    such a prestigious commission, but it can be fairly speculated that a certain amount
    of dissatisfaction surfaced in their dealings with Lelli.

81 Archivio della Accademia delle Belle Arti di Bologna, *Atti Accademia Clementina*,
    MS I, fols. 65–66. Cited by Focaccia in her *Anna Morandi Manzolini*, p. 40, note
    148.

82 See Medici, "Elogio di Giovanni Manzolini e di Anna Morandi, coniugi Manzo-
    lini," in *Memorie della Accademia delle Scienze dell'Istituto di Bologna* (Bologna: San
    Tommaso d'Aquino), 8:4–5.

83 Ibid.

84 See Fantuzzi, *Notizie degli scrittori bolognesi*, 5:50–51; BCAB, MS B 134, vol. 12,
    Oretti, *Notizie de' Professori*, fols. 120–23; and Crespi, *Felsina Pittrice*, 301–7. Among
    recent historians to assert a clear position on the controversy are Riccomini,
    Mostra della scultura bolognese del Settecento, 109; Ferretti, "Il nomista e il canonico";
    Fanti, "Sulla figura e l'opera di Marcello Oretti"; and Boschloo, *L'Accademia
    Clementina*.

85 BUB, MS 243, Letter of Beccari to Scarselli, Bologna, 18 September 1749. Miriam
    Focaccia cites this correspondence in her introduction to *Anna Morandi Manzolini*,
    41–43.

86 Ibid.

87 BUB, MS 3882, caps. LVIII, A 5, *Cinque lettere da Roma di Marco Antonio Laurenti a
    Ercole Lelli* (1747–53), 9 October 1747. Laurenti's denial notwithstanding, the pope

had doubtless heard about Morandi from his many correspondents, but he had not yet given her his support.

Chapter 2

1 *Le Costituzioni dell'Istituto delle Scienze. Eretto in Bologna sotto li 12 Dicembre 1711*. In Angelini, *L'Istituto delle Scienze e L'Accademia*, vol. 3 of *Anatomie Accademiche*, chap. 5, article 3, p. 510.

2 BUB, MS 2193, Morandi, *Anatomical Notebook*, fols. 72r–78r.

3 Ibid., fol. 75v.

4 ASB, Archivio Ranuzzi, *Istrumenti scritture diverse spettanti alla Nobile Casa Ranuzzi dall'anno 1769 all'anno 1773*, Libro 124, n. 21, "Nota degli strumenti per anatomia venduti dalla Signora Manzolini al Sign.r Sen.re Ranuzzi."

5 My sincere thanks to the science historian Mike Shank for explaining that this unit, being close to 1–2 mm, was likely the equivalent unit of measure as the *punti* used by Galileo in his experiments.

6 For references to the necessities of fire and light for the home anatomy studio, see ASB, Assunteria di Studio, *Diversorum*, busta 91, "Relazione degli Assunti ed altra circa gli Anatomici gli'interesse egli […] dell'Anatomia," May 1755.

7 My thanks to Marta Cavazza for indicating the Dogaressa's interest in human anatomy. Unfortunately, requests to view the Dogaressa's archived letters at the Palazzo Mocenigo have thus far been refused.

8 Pompenius Gauricus, *De sculptura* (Florence, 1504).

9 Pompenius Gauricus, *De sculptura*, trans. into German and with an introduction by H. Brockhaus (Leipzig, 1886), p. 128; cited in Phoebe Dent Weil, "Bozzetti Problems: Philological, Functional Technical," Ph.D. diss., New York University, 1966. My thanks to the author and art restorer for sharing her work on this subject with me.

10 Giorgio Vasari, *Vite de' più eccellenti architetti, pittori, et scultori italiani*, ed. Gaetano Milanesi (Firenze: G. C. Sansoni, 1878), 1:152.

11 See Francesco Maria Zanotti, "De re obstetrica," under the rubric "Medica," in *De Bononiensi Scientiarum et Artium Instituto atque academia commentarii* (Bologna: Lelio della Volpe, 1755), 3:87–89; and Luigi Galvani, "De Manzoliniana supellectili oratio habita in Scientiarum et Artium Instituto cum ad anatomen in tabulis ab Anna Manzolina . . . ," trans. in Italian by Luciana Quadrelli in *Alma Mater Studiorum*, 94–103.

12 My thanks to Dr. Marta Poggesi, General Coordinator for the Specola Museum of Natural History in Florence, for explaining the technique of wax modeling and demonstrating the workings of wooden and clay casts.

13 Angelini, *L'Istituto delle Scienze e l'Accademia*, 3:60–63.

14 "Imparai a conoscere i libri più scielti, ed a suo tempo il metodo di medicare." Marcello Malpighi, *Memorie di me Marcello Malpighi ai miei posteri fatte in villa l'anno 1689* (Bologna: Zanichelli, 1902). Cited by the most important Malpighi translator and scholar, Howard B. Adelmann, in his oration at Pisa, 1972; online at http://www3.humnet.unipi.it/galileo/fondazione/Vincitori%20Premio%20Galilei/Howard_Adelmann.htm

and published compositions, which are dull enough, I should suspect that this impromptu-exercise seldom leads to poetical excellence.

Joseph Forsyth, *Remarks on Antiquities, Arts and Letters during an Excursion in Italy in the Years 1802 and 1803* (London: John Murray, 1824), 61.

10 F. Zanotti, "De re obstetrica," 3:87–89. Translated from the Latin by George Pepe.

11 Crespi, *Felsina pittrice*, 307.

12 BUB, MS 243, Iacopo Bartolomeo Beccari, *Lettere autografe al Dott. Flaminio Scarselli (1742–1761)*, Bologna, 18 September 1749. Quoted in Focaccia, "Anna Morandi Manzolini," 41.

13 Giuli, "Corilla Olimpica," 165.

14 The following biographers liberally reiterate Crespi's characterization of Morandi: Fantuzzi, *Notizie degli scrittori bolognesi*, 6:113–16; Carolina Bonafede, *Cenni biografici e ritratti d'insigni donne bolognesi* (Bologna: Sassi, 1845), 166–72; Giuseppe Bosi, *Archivio patrio di antiche e moderne rimembranze Felsinee* (Bologna: Tipografia di Antonio Chierci, 1855), 241–45; Medici, "Elogio di Giovanni Manzolini e di Anna Morandi," 8:3–23; also by Medici, *Compendio storico*, 356–62.

15 F. Ruggeri, "Il Museo dell'Istituto di Anatomia Umana Normale"; and Vittoria Ottani and Gabriella Giuliani-Piccari, "L'opera di Anna Morandi Manzolini nella ceroplastica anatomica bolognese," in *Alma Mater Studiorum*, 81–93.

16 Dacome, "Waxworks and the Performance of Anatomy," esp. 32–33. "As her hand holding the now lost scalpel could easily be mistaken for a sewing hand, the portrait also represented the anatomist in a gesture that associated her work with the largely feminine world of embroidery."

17 Crespi not only vigorously defended Giovanni Manzolini as the authentic artist and anatomist of the papal commission in sketches of Manzolini and Morandi, he omitted any mention of Lelli from his *Lives of Bolognese Artists* (*Felsina pittrice: Vite de' pittori bolognesi*).

18 Crespi, "Anna Morandi Manzolini," in *Felsina pittrice*, 309.

19 Although a number of their contemporaries recount that Lelli kept Crespi out of the Clementina, no documents have thus far surfaced to substantiate this claim.

20 Giovanna Perini, "La Camera Anatomica dell'Istituto delle Scienze," in *Palazzo Poggi da dimora aristocratica a sede dell'università*, ed. Anna O. Calvina (Bologna: Nuova Alfa Editoriale, 1988), 177. Perini has encapsulated the dominant view of the quarrel that divided the Clementina for a generation. It was based, she maintains,

> in practice on two antithetical concepts of artistic activity seen as, on the one side, etiologically and teleologically autonomous and, on the other, as heteronymous. This latter interpretation upheld and propagated by friends and students of Lelli, among whom Francesco Algarotti, Carlo Pisarri or Carlo Bianconi, will have an international irradiation. . . . The autonomous version, ideologically elitist of an art hierarchically distinct in minor and major branches, will instead have its greatest champion outside of the Academy in Luigi Crespi. (Ibid.)

Crespi, in his assault on Lelli, thus represents for Perini the voice of greatest opposition to a new union between art and science. As we will see, this does not explain why Crespi praises Manzolini effusively in his *Lives of Bolognese Artists*. Massimo Ferretti also describes the quarrel as a contest between an old and new cultural aesthetic: "Il notomista e il canonico," 104.

21 Crespi, *Felsina pittrice*, 160.

22 Crespi does not name Lelli in his attack on anatomical studies, but it is clear that Lelli is his target.

23 Contemporary scholars still engaged in this debate include Ferretti on the side of Lelli in "Il notomista e il canonico," 110–14; again in support of Lelli, Boschloo, *L'Accademia Clementina*, 49–99; and offering a compelling argument against the competence and integrity of Lelli is Fanti, "Sulla figura e l'opera di Marcello Oretti," 125–43.

24 Crespi, "Giovanni Manzolini," in *Felsina pittrice*, 301.

25 Ferretti, Fanti, and Boschloo offer no consideration of Morandi's position within this quarrel.

26 BUB, MS 3882, caps. LVIII A 5, *Cinque lettere da Roma di Marco Antonio Laurenti a Ercole Lelli* (1747–53), 9 October 1747.

27 For a detailed history of the collection, see *Ars obstetricia bononiensis: Catalogo ed inventario del Museo Obstetrico Giovan Antonio Galli* (Bologna: CLUEB, 1988). Manzolini and Morandi were hired to sculpt only the first twenty models, mainly because of cost. The commission passed to Giambattista Sandri, who created the bulk of the figures in the less expensive material of clay.

28 On Galli's now-lost drawings, see Marco Bortolotti, "Il maestro alla lavagna: Il museo del Galli dall'inventario al catalogo," in *Ars obstetricia bononiensis*, 22. François Mauriceau, *Description anatomique des parties de la femme . . .* (A Leide, Chez la Veve de Bastiaan Schouten, 1708); François Mauriceau, *Traité des maladies des femmes grosses, et de celles qui sont nouvellement accouchées* (Paris, 1668); Hendrick van Deventer, *Observations sur la grossesse et l'accouchement des femmes, et sur leurs maladies & celles des enfans nouveau-nez . . .* (Paris: L'Auteur, 1695), and Hendrick van Deventer, *Operationes chirurgicae novum lumen exhibentes obstetricantibus . . .* (Lugduni Batavorum: Apud Andream Dyckhuisen, 1701). In Oretti's words, "[Anna Morandi] made for Dr. Galli a great series of uteruses and fetuses admirably formed that later passed to a room on the bottom floor of the Institute and that continue to serve a school of scholars of Medicine that in this great Mother of Learning exercize their talent." BCAB, MS B 133, vol. 11, Oretti, *Notizie de professori*, fol. 227.

29 Pancino, "L'ostetricia del Settecento," 24. Pancino clarifies that women who practiced midwifery were typically older women, often widows, who had experienced motherhood themselves.

30 Ibid., 26.

31 On 22 November 1751, a group of senators visited Galli at his home studio, where they were awed by his wax and clay models of the developing fetus and the machine he had created to instruct midwives in proper birthing techniques. Bortolotti, "Il maestro alla lavagna," 18.

32 On 20 and 24 August 1757, Galli wrote to Flaminio Scarselli to lament Lelli's interference with the establishment of his obstetrical museum in the Institute of Sciences. Galli maintained that Lelli was jealous of the superior waxes created by Manzolini and Morandi. BUB, MS 72, Carteggio Scarselli, *Lettere a Flaminio Scarselli*, vol. 1 (1742–60), pp. 98–109.

33 On the incursion of male medical professionals into women's traditional domain of childbirth and delivery, see Adrian Wilson, *The Making of Man Midwifery: Childbirth in England 1660–1770* (Cambridge, MA: Harvard University Press, 1995).

34 Giovan Antonio Galli, "Proposal for the Reform of Obstetrical Practice in Bologna," ASB, *Diversorum*, vol. 8, n. 3; cited by Viviana Lanzarini, "Un museo per la didattica e la sanità ostetrica," in *Ars obstetricia bononiensis*, 42.

35 On Madame du Coudray, see the wonderful biography by Nina Rattner Gelbart, *The King's Midwife* (Berkeley and Los Angeles: University of California Press, 1998).

36 These are identified in the inventory as figures 156, 157, and 158. See *Ars obstetricia bononiensis*, 94–95.

37 On the theories of Deventer and his contemporaries regarding the placenta, see Walter Radcliffe, *"Milestones in Midwifery" and "The Secret Instrument (The Birth of the Midwifery Forceps)"* (San Francisco: Norman Publishing, 1989), 39–60.

38 See articles by Bortolotti, Pancino, and Lanzarini in *Ars obstetricia bononiensis*, 14–42.

39 BCAB, MS B 134, vol. 12, Oretti, *Notizie de' professori*, "Giovanni Manzolini," fol. 135. Crespi, *Felsina pittrice*, 306–7. The pope established the courses in Italy on the "the demonstration of surgical operations on cadavers," with Molinelli as the head on 23 August 1742. Benedictus Papa XIV, "Motu proprio per il quale si istituisce in Bologna una scuola di chirurgia," in *Lettere, brevi, chirografi, bolle ed appostoliche determinazioni 1740–55*, 3 vols. (Bologna, 1749–56), 1:258.

40 Josias Weitbrecht (1702–1747), member and anatomy and physiology lecturer of the Imperial Academy of Science at St. Petersburg from 1725, was perhaps best known as author of a comprehensive anatomy of the joints, *Syndesmolgia sive historia ligamentorum corporis humani . . .* (St. Petersburg: Typographia academiae scientiarum, 1742). The reference in Medici (see note 41 below) no doubt concerns his catalogue of anatomical preparations in the St. Petersburg Chamber of Arts that Emperor Peter I had bought from Frederick Ruysch in 1717. Johann Georg Roederer (1726–63), professor of obstetrics at the University of Göttingen and author of *Elementa artis obstetriciae in usum auditorum denuo edidit, nec non praefatione et annotationibus instruxit H. A. Wrisberg* (Göttingen: Albizziniana, 1759), was a renowned German obstetrician who contributed to the anatomy and physiology of the fetus and the physical process of giving birth. My sincere thanks to Lilla Verkerdy for her help in identifying Weitbrecht, who is cited in Medici as Weibrochio.

41 Medici, "Elogio di Giovanni Manzolini e di Anna Morandi," 8:6.

42 Focaccia, "Anna Morandi Manzolini," 41–42.

43 Luigi Belloni, "Suono e orecchio dal Galilei al Valsalva: Nel Terzo Centenario della nascita di A. M. Valsalva," *Simposi Clinici CIBA* 3, no. 3 (1966): 33–42.

44 The tracts have been reproduced in R. A. Bernabeo and I. Romanelli, "Considerazioni di Giovanni Manzolini (1700–1755) sull'anatomia dell'orecchio in condizioni normali e patologiche," in *Atti del XXVII Congresso Nazionale di Storia della Medicina* (Caserta, Capua-Salerno, Italy, September 1975). This citation is found on one of several unnumbered pages at the center of the essay, where Manzolini's original tracts are reproduced.

45 Ibid.

46 Ibid.

47 To clarify, Jacopo Beccari was president of the Academy of Science within the Institute of Sciences in 1723, 1735, 1740, and 1749. He was president of the institute from 1750 to 1766.

48 She seems to be referring here to the internal acoustic meatus that harbors the facial nerve and vestibulocochlear nerve, which she considered a single bifurcated nerve.

49 The semicircular ducts or membranous labyrinth is described here.

50 She is describing the facial nerve, whose distinct attachments she was unable to distinguish from those of the vestibulocochlear nerve.

51 This is the base of the modiolus.

52 Curiously, she gets this wrong as this, the facial nerve, is visibly larger than the vestibulocochlear nerve.

53 BUB, MS 2193, Morandi, *Anatomical Notebook*, fol. 21r–v. Infinite thanks go to Professor of Anatomy Krikor Dikranian for his assistance in deciphering the anatomical structures to which Morandi was referring.

54 Here he is referring to the facial nerve, which he was unable to distinguish from the vestibulocochlear nerve, and the branches he describes appear to be the chorda tympani greater petrosal nerve.

55 He means the branches of the vestibular nerve.

56 Galvani, "De Manzoliniana," 100.

57 Under the direction of his father—Augustus Quirinus Rivinus (1652–1723), a well-known professor of physiology, botany, and medicine of the University of Leipzig—Johannes Augustus Quirinus Rivinus wrote a tract on the ear, *De auditus vitiis* (Leipzig, 1717), in which the theory of the "incisura tympanica" was introduced. My thanks to Lilla Verkerdy for her help in identifying references to Rivinus's work. On this theory, see Adrian Griffith, "The Foramen of Rivinus—An Artefact," in *Journal of Laryngology and Otology* (July 1961): 607–13.

58 BUB, MS 2193, Morandi, *Anatomical Notebook*, fol. 20 r–v.

59 Crespi, *Felsina pittrice*, 303; BCAB, MS B 134, vol. 12, Oretti, *Notizie de' professori*, fol. 135.

60 See letters to Bianchi by Morandi of 15 April and 24 May 1755, Biblioteca Civica di Gambalunghiana, Rimini, Fondo Gambetti, *Lettere autografe al Dott. Giovanni Bianchi*, fasc. Manzolini Morandi, Anna e Giovanni Manzolini, 15 April 1755.

61 These texts and many others are cited in Giorgio Cusatelli, ed., *Viaggi e viaggiatori del Settecento in Emilia e in Romagna* (Bologna: Il Mulino, 1986), 2:596–601.

62 Carl Jörg, *Pfalzgraf Friedrich Michael von Zweibrüken und das Tagebuch seiner Reise nach Italien* (Munich, 1892), 76–79. "Die anatomie in Wacchs ist eines der feinsten Kunststücke un wurde von einer Frau allda sehr gelehrt und beredtsam expliziert."

63 My sincere thanks to Matthew Erlin for translating this passage.

64 She is undoubtedly the count's guide, as no other woman would have approached her fame and expertise in the specialty of anatomical ceroplasty.

65 See Laurenti's letter to Lelli soothing his anxiety over Morandi's rise: BUB, MS 3882, caps. LVIII A 5, *Cinque lettere da Roma di Marco Antonio Laurenti a Ercole Lelli* (1747–53), 9 October 1747.

66 "J'ai été aussi voir à Bologne avec M. Wolff une femme qui fait en cire avec la dernière perfection toutes le parties du corps humain. Elle donne des leçons d'anatomie à de jeunes élèves avec une nouvelle mèthode et des connaissances

admirables. Elle nous a dit avoir aussi diséqué mille cadavres dans sa vie et il ne se passe pas de mois que les hôpitaux de Bologne ne lui en fournissent quelques uns. Nous sommes sortis de chez elle tout pénétrés d'admiration pour ses divins talents." Vicomte Ch. Terlinden, "Journal de voyage d'un médecin bruxellois de Munich à Rome en 1755," *Bulletin de l'Institut historique belge de Rome* 23 (1944–46): 135. My thanks to Elizabeth Allen for her help with the translation of this passage.

67 The eighteenth-century connotations of this term are the subject of chapter 4.

68 Cavazza, *Settecento inquieto*, 63.

69 On Giovanni Bianchi, see Stefano de Carlis and Angelo Turchini, *Giovanni Bianchi: Medico primario di Rimini ed archiatra pontificio* (Villa Verrucchio: Pazzini, 1999).

70 Giovanni Bianchi, "Letter Written from Bologna," 708–11.

71 Ibid.

72 Dacome offers a different interpretation of this episode in the more general and retrograde terms of a correspondence between maternal and artistic creativity and between the impression of the maternal imagination on the fetal body and impressing a form on wax. "Women, Wax and Anatomy," 529–41.

73 Galileo Galilei, *Dialogue concerning Two Chief World Systems, Ptolemaic and Copernican*, trans. Stillman Drake (Berkeley and Los Angeles: University of California Press, 1967), 51.

74 Park, *Secrets of Women*, 26–27.

75 Sawday, *The Body Emblazoned*, 1–15.

76 Biblioteca Civica di Gambalunghiana, Fondo Gambetti, *Lettere autografe al Dottor Bianchi. Fascicolo Manzolini Morandi Anna e Manzolini Giovanni*. The letters were signed by both Anna Morandi and her husband, Giovanni Manzolini; however, there is ample evidence that Morandi composed them. Her signature matches the handwriting of the letters and appears first in each. Moreover, her husband died of a debilitating illness three months after the first letter was sent.

77 It seems that Bianchi himself did not know the identity of the anatomists he denounces in his letter.

78 Morandi was not the only scientific woman correspondent of Bianchi's. For an intriguing account of Bianchi's epistolary exchange with the Cartesian philosopher Laura Bentivoglio Davia, see Paula Findlen, "Women on the Verge of Science: Aristocratic Women and Knowledge in Early Eighteenth-Century Italy," in *Women, Gender and Enlightenment*, ed. Sarah Knott and Barbara Taylor (London: Palgrave Press, 2005), 265–87.

79 Giovanni Bianchi, *Lettera del Signor Dottore Giovanni Bianchi medico primario d'Arimino ad un suo amico di Cesena . . .* (Armino: Stamperia Albertiniana, 1755).

80 References to the case appeared in the March 1750 issue of the *Novelle letterarie*; in the *Acts* of the Lissia in 1751, p. 709; again in *Novelle letterarie* col. 833, in 1751; again in the *Atti di Lissia* in 1752, p. 214; and in many other publications across Italy. It even appeared in the January 1753 issue of the Parisian *Journal de Trévoux*, p. 92. Indeed, Giovanni Lami, the editor of the journal *Novelle letterarie* (1740–70), remarked that Bianchi's extensive publications on the case made it known not only across Italy but all of Europe, along the "shores of the Arno, the Tiber, the Danube and the Thames": Giovanni Lami, *Novelle letterarie pubblicate a Firenze* (Florence: Marco Lastri, 1755), 395.

81 Bianchi's theories on the distinct influence of the cerebrum and the cerebellum on the lower body were apparently not as original as he imagined, however, and

Morgagni responded to the "case history" with anemic admiration: "Your Diligence in reobserving attentively that which many other Anatomists have observed but have not described with the same exactitude seems to me deserving of praise." Morgagni, letter to Bianchi, "A me parve degna di lode la Diligenza di Lei in riosservare attentamente ciò che tanti altri Notomisti e osservando, non avevano con pari esattezza descritto." In Giovanni B. Morgagni, *Carteggio inedito di G. B. Morgagni con Giovanni Bianchi*, ed. Guglielmo Bilancioni (Bari: Società Tipografica Editrice Barese, 1914), 195.

82 Lami, *Novelle letterarie pubblicate a Firenze*, 393. On Lami and the focus of the journal, see the biographical sketch "Giovanni Lami," in Sandra Baldacci and Valerio Bartoloni, *Giovanni Lami e il Valdarno inferiore*, ed. Valerio Bartoloni (Pisa: Pacini Editore, 1997), 15–22.

83 *Lettera del Signor Dottor Giovanni Bianchi Medico Primario d'Arimino ad un suo amico di Cesena sopra un preteso supplimento alla storia medica d'una postema del lobo destro del Cerebello pubblicato dal Sig. Dottor Carlo Serra della medesima città* (Arimino, Italy: Stamperia Albertiana, 1755), 17–19.

84 Domenico Gusmano Galeazzi (1686–1775) was professor of anatomy and physics, among whose students was the electrophysiologist Luigi Galvani. Tommaso Laghi (1709–1764) was an influential anatomist at the university and an avid critic of the Hallerian theory of muscular motion. Jacopo Bartolomeo Beccari (1682–1766) was professor of physics and among the first professors of chemistry at the university. As we have seen, Giovanni Antonio Galli (1702–1782) was a surgeon, physician, and founder of Bologna's first school of obstetrics, in which he used wax facsimiles of the female reproductive system. Seen previously, Pier Paolo Molinelli (1702–1764) was the first chair of operative medicine in Bologna. Francesco Maria Galli Bibiena (1720–1774) was a student of Jacopo Beccari's who studied medicine at Bologna and became a university lecturer in the same subject.

85 Biblioteca Civica di Gambalunghiana, Fondo Gambetti, *Lettere autografe a Dott. Giovanni Bianchi*, letter to Bianchi from Anna Morandi Manzolini and Giovanni Manzolini, 7 April 1755.

86 Ibid.

87 Ibid.

88 Biblioteca Civica di Gambalunghiana, Fondo Gambetti, *Lettere autografe*, 24 May 1755.

89 Ibid.

Chapter 4

1 Giuseppa Eleonora Barbapiccola, "La traduttrice a' lettori," in *I Principi della filosofia di Renato Des-cartes tradotti dal francese col confronto del latino in cui l'autore gli scrisse* (Turin, 1722), trans. Paula Findlen in *The Contest for Knowledge: Debates Over Women's Learning in Eighteenth-Century Italy*, ed. Rebecca Messbarger and Paula Findlen (Chicago: University of Chicago Press, 2005), 48.

2 Notably, in the 2000 restoration of her works, a receipt with her signature was found, tucked inside the satin bodice of Morandi's self-portrait, for a charitable contribution she had given a local church—a small act of contrition, perhaps, for embellishing her image and her importance. My thanks to Maricetta Parlatore for showing this to me.

3 Mary Sheriff, *The Exceptional Woman: Elisabeth Vigée Lebrun and the Cultural Politics of Art* (Chicago: University of Chicago Press, 1996), esp. 180–220.

4 Perini, "La Camera Anatomica dell'Istituto delle Scienze," 176–88.

5 My thanks to Luciano Guerci for providing me with the following documents that attest to Morandi's affiliation with Caldani and Manfredi: Letter of Ignazio Somis to Eraclito Manfredi, 29 March 1758, BCAB, MS Collezione degli autografi LXV, 17.584–17.754 numero progressivo; 17.710, copia autografa; Letter of Somis to Caldani, 12 April 1758, BCAB, MS Collezione degli autografi LXV, 17.584–17.754 numero progressivo; 17.712, copia autografa.

6 Albrecht von Haller, ed., *Praelectiones academicae in proprias institutiones rei medicae*, by Herman Boerhaave (Gottingen: A. Vandenhoeck, Gottingen, 1739), 129. Cited by Shirley Roe, *Matter, Life, and Generation: 18th-Century Embryology and the Haller-Wolff Debate* (Cambridge: Cambridge University Press, 1981), 33. On this controversy, see Cavazza, "La recezione della teoria halleriana."

7 Morandi's archival collection is listed and described by Raffaele A. Bernabeo, "La libreria scientifica di Anna Morandi Manzolini."

8 Carlino, *Books of the Body*, 8–68.

9 The term refers to the formalized performance of a scholastic debate, exercise, or disputation, typically on philosophical or theological questions. Carlino borrows the term from Heckscher's important study, *Rembrandt's Anatomy of Dr. Nicolaas Tulp*, 45–46. See n. 9 in Carlino, *Books of the Body*, 13.

10 Ibid., 51.

11 Ibid., 51–53.

12 Porter, "Medical Science and Human Science in the Enlightenment," 56.

13 ASB, Archivio Ranuzzi, *Istrumenti scritture*, Libro 124, n. 21, "Nota degli Strumenti per Anatomia venduti dalla Signora Manzolini al Sign.r Sen.re Ranuzzi."

14 Benedict XIV, *De servorum*, 4:461–67. The full citation reads, "It is deduced from what has been said that the fantasy or imagination is none other than that admirable book of the brain in which are printed both intellectual notions as well as images of perceptible objects collected by the senses." On Benedict XIV's engagement with contemporary medical and philosophical theories, see Dacome, "Women, Wax and Anatomy," 537–41.

15 On this analogy, see Park, *Secrets of Women*, 66.

16 AAB, *Libri mortuorum*, vol. III, 8 June 1755, fol. 42.

17 ASB, Gabella Grossa, *Atti delle Congregazioni I/42*, 15 December 1755, fol. 112 r–v.

18 Archivio Accademia Belle Arti di Bologna, *Atti dell'Accademia Clementina*, vol. 1, fol. 190, 3 December 1755.

19 Only men are listed in the Acts of the Academy dating from 1710 until 1804, while women's names appear solely as honorary members. After Morandi's admission and until the end of the century, fourteen other native and foreign honorary female members would be installed. My sincere thanks to Prof. Antonietta De Fazio for her generous help with relevant archival information.

20 Anna Piattole, Florentine painter, inducted 10 April 1760; Ortensia Poncarali, 30 August 1761; Rosalba Bisacck Bruni, illustrator and embroiderer, 16 October 1763; Dorotea Therbouc di Lisiewska, painter trained in Germany, 22 August 1763; Teresa Orsini di Cassine, miniaturist from Alessandria, Italy, 15 June 1765; Maddalena Morelli Fernandez, the improvisational poet known as Corilla Olimpica

from Pistoia, 8 April 1766; Camilla Bobbadiferro (Boccaferro) Marulli, 6 March 1767; Girolama Boccadiferro Legnami, 16 March 1767; Eleonora Monti, Bolognese painter, 9 November 1767; Teresa Martelli Casali, miniaturist, 14 April 1768; Rose Matthieu, painter, 26 January 1770; Joanna Juliana Fiederike Bacciarelli, figure painter, 7 April 1771. On the history of Bolognese women artists, see Laura Ragg, *The Women Artists of Bologna* (London: Methuen and Co., 1907); Fredrika H. Jacobs, *Defining the Renaissance Virtuosa* (Cambridge: Cambridge University Press, 1997); and Caroline P. Murphy, *Lavinia Fontana: A Painter and Her Patrons in Sixteenth-Century Bologna* (New Haven, CT: Yale University Press, 2003).

21 In her notebook, each section begins with this declaration of her membership in the institute.

22 ASB, Assunteria d'Istituto, *Diversorum*, busta 10, no. 5, "Compere dal Conte Girolamo Ranuzzi dello Studio e delle preparazioni anatomiche di Anna Manzolini e notizie attinenti."

23 BCAB, MS Fondo Mondini, *Richiesta all'Ospedale di S. Maria della Morte dell'Amm. re C.o Vittori per procurar cadaveri alla Anna Manzolini*, Cart. VIII, n. 3 (1756).

24 Cavazza, "Dottrici e lettrici," 118.

25 ASB, *Orfanotrofio di S. Maria Maddalena e di S. Bartolomeo uniti*: "Campione dei puti," fol. 26.

26 The placement of children in orphanages for adoption was not a rarity in the period. Although there is no indication of why Giuseppe was chosen by the Solimei family, it is likely that the fame of his parents, especially his mother, played some part in his adoption. See Nicholas Terpstra, *Abandoned Children of the Italian Renaissance: Orphan Care in Florence and Bologna* (Baltimore: Johns Hopkins University Press, 2005).

27 It must be noted that at about this same time in France, Marie Marguerite de Biheron (1719–1795) was gaining renown as an anatomist and anatomical wax modeler, especially for her obstetrical models, although she illustrated in wax nearly all parts of the body. She left no writings on her anatomical practice, however. It is not known if she and Morandi knew of each other, although they received visits to their respective studios from a number of the same dignitaries. On Biheron, see Londa Schiebinger, *The Mind Has No Sex? Women in the Origins of Modern Science* (Cambridge, MA: Harvard University Press, 1989), 27–29.

28 BCAB, MS B 133, vol. 11, Oretti, *Notizie de professori*, fol. 229.

29 It was sold to Ranuzzi in 1769 for 500 *zecchini*, while the most paid for any other series in Morandi's collection was 50 *zecchini*.

30 This was discussed at length in chapter 2.

31 Crespi, *Felsina pittrice*, 309–13: BCAB, MS B133, vol. 11, Oretti, *Notizie de professori*, fols. 227–30; Fantuzzi, *Notizie degli scrittori bolognesi*, 6:113–16.

32 Archivio di Stato, Firenze, Accademia del disegno. *Primi compagni di pittori*, no. 121, 112, 21.

33 On Maria Theresa's reforms, see Richard Schober, "Gli effetti delle riforme di Maria Teresa sulla Lombardia," in *Economia, istituzioni e cultura in Lombardia nell'età di Maria Teresa*, ed. Aldo de Maddalena, Ettore Rotelli, and Germano Barbarisi (Milan: Il Mulino, 1982), 3:201–14.

34 BCAB, *Regolamenti per gli esercizi letterarj dell'Accademia de' Gelati Restituiti nella sessione del 31 luglio 1786*, p. 6. On the academy, which was founded in 1588, see

Fantuzzi, *Notizie degli scrittori bolognesi*, 1:11–13.

35 No records have been found to verify the often cited commissions and invita-
    tions from the Royal Society. Chapter 7 will present sizable primary evidence of
    Catherine the Great's avid and prolonged interest in the work of Anna Morandi.

36 BCAB, MS Collezione degli autografi LXV, numero progressivo 17.692; 17.712,
    17.710, cc 1735–88, copia autografa.

37 ASB, Assunteria d'Istituto, *Atti*, vol. 5 (1754–60), 4 January 1759.

38 Jean Astruc, *Traité des maladies des femmes*, 6 vols. (Paris, 1761–65).

39 Germano Azzoguidi, *Observationes ad uteri constructionem pertinentes* (Bologna:
    Longhi, 1773), 36–37. My thanks to George Pepe for his translation of this
    passage.

Chapter 5

1 G. J. Barker-Benfield, *The Culture of Sensibility: Sex and Society in Eighteenth-Century
   Britain* (Chicago: University of Chicago Press, 1992), xvii–xviii.

2 Gasparo Gozzi, *Gazzetta veneta* [1762], ed. Antonio Zardo (Florence: Sansoni,
   1967), 125.

3 Antonio Conti, "Lettera dell'Abate Conti P. V. sopra lo stesso argomento,"
   appendix to Antoine Léonard Thomas, *Saggio sopra il carattere i costumi e lo spirito delle
   donne ne' varii secoli*, ed. Lodovico Antonio Loschi, (Venice: Giovanni Vitto, 1773),
   224. On this text and eighteenth-century scientific misogyny, see Messbarger,
   The Century of Women, 49–68. See also Ludmilla Jordanova, "Sex and Gender," in
   Inventing Human Science, 152–83. Primary arguments on the subject also include
   Giovanni Antonio Volpi, "Che non debbono ammettersi le donne allo Studio
   delle Scienze, e delle belle arti . . . ," in *Discorsi accademici di vari autori viventi intorno
   agli studi delle donne*, ed. Giovanni Antonio Volpi (Padua: Giovanni Manfrè, 1729),
   27–45; and Ferdinando Galiani, "Croquis d'un dialogue sur les femmes" [1772],
   in *Illuministi italiani*, vol. 6, *Opere di Ferdinando Galiani*, ed. Furio Diaz and Luciano
   Guerci (Milan: Riccardo Ricciardi, 1975), 615–33.

4 Antonio Conti, *Prose e poesie del signor abate Antonio Conti, patrizio veneto* (Venice:
   Giambattista Pasquali, 1756), 2:lxv–lxxv.

5 Galiani, "Croquis d'un dialogue sur les femmes."

6 Notably, Morandi's name surfaces in a number of tracts written in defense of
   women and the female intellect. For example, in the anonymous "Defense of
   Women," published in Milan in 1767 at the height of Morandi's career, a footnote
   reads: "Una bolognese moglie d'un fabbricatore di statue di cera, perfezionó
   quest'arte" (The Bolognese wife of a craftsman of wax statues perfected this art),
   p. 13, n. 2. My thanks to Paula Findlen for informing me about this text.

7 See Messbarger, "Waxing Poetic: Anna Morandi Manzolini's Anatomical
   Sculptures," *Configurations* 9 (2001): 65–97, and "Cognizione corporale: La poetica
   anatomica di Anna Morandi Manzolini," in *Scienza a due voci*, ed. Raffaella Simili
   (Florence: Olschki, 2006), 39–61.

8 Messbarger, "Re-membering a Body of Work: Anatomist and Anatomical
   Designer Anna Morandi Manzolini," *Studies in Eighteenth-Century Culture* 32
   (2003): 130, and "Re-casting Wax Anatomical Modeler Anna Morandi Manzolini
   (1714–1774)," in *Dall'origine dei lumi alla rivoluzione: Scritti in onore di Luciano Guerci*

e Giuseppe Ricuperati, ed. Donatella Balani, Dino Carpanetto, and Marina Roggero (Rome: Edizioni di Storia e Letteratura, 2008), 353–84.

9   See Krzysztof Pomian, "Vision and Cognition," in *Picturing Science Producing Art*, ed. Caroline A. Jones and Peter Galison (New York: Routledge, 1998), 211–32.

10 Findlen, "Science as a Career in Enlightenment Italy," 452.

11 Ibid. On Bassi's early Newtonianism, see Cavazza, "Laura Bassi," and Massimo Mazzotti, "Newton for Ladies," Bologna Science Classics Online: http://www. cis.unibo.it/cis13b/bsc03/notbyed2.ASP?id_opera=32&offset=1.

12 Mazzotti, "Newton for Ladies," pp. 4–6.

13 Angelini, *L'Istituto delle Scienze e l'Accademia*, 329.

14 See G. Martinotti, *L'Insegnamento dell'anatomia in Bologna Prima del Secolo XIX* (Bologna: Azzoguidi, 1911), 131–33.

15 Isaac Newton's own terms from the *Opticks* (New York: Courier Dover, 1979), 63.

16 BUB, MS 2193, Morandi, *Anatomical Notebook*, fol. 5r.

17 Ibid., fol. 10r.

18 Ibid., fol. 14r.

19 Ibid., fol. 14v.

20 Ibid., fol. 13v.

21 Ibid., fol. 7r–v.

22 On the role and representation of the hand as a means and a mnemonic of knowledge, see the important essays in the recent volume edited by Claire Richter Sherman, *Writing on Hands: Memory and Knowledge in Early Modern Europe* (Carlisle, PA: The Trout Gallery, Dickinson College, 2001).

23 BUB, MS 2193, Morandi, *Anatomical Notebook*, fol. 50 r–v.

24 Ibid., fol. 50v.

25 Ibid., fol. 51r.

26 John Locke's monumental *Essay concerning Human Understanding*, of course, had a profound influence on subsequent Enlightenment sensism. On pleasure and pain, he states in book 2, chapter 20:

>   Amongst the simple ideas which we receive both from sensation and reflec-
>   tion, pain and pleasure are two very considerable ones. For as in the body
>   there is sensation barely in itself, or accompanied with pain or pleasure, so
>   the thought or perception of the mind is simply so, or else accompanied also
>   with pleasure or pain, delight or trouble, call it how you please. These, like
>   other simple ideas, cannot be described, nor their names defined; the way of
>   knowing them is, as of the simple ideas of the senses, only by experience.

Locke's "Essay concerning Human Understanding": Books II and IV (with Omissions), ed. Mary Whiton Calkins (London: Open Court Publishing, 1905), 121–22.

27 This term is from Italo Calvino, *Six Memos for the New Millennium* (Cambridge, MA: Harvard University Press, 1988), 82.

28 BUB, MS 2193, Morandi, *Anatomical Notebook*, fol. 52v.

29 Ibid., fol. 58r.

30 Francesco Algarotti, *Il Newtonianismo per le dame ovvero dialoghi sopra la luce i colori e l'attrazione* (Naples: Giambattista Pasquali, 1739), 89–103.

31 Ibid., 90–92.

32 Nancy Siraisi, *Medicine and the Italian Universities 1250–1600* (Leiden: Brill, 2001), 257–62.

33 Heckscher, *Rembrandt's Anatomy of Dr. Nicholas Tulp*, 73.

34 Heckscher asks this same question regarding these cited anatomists, whose fame obtained from their expertise in other parts of the body: ibid., 73–74.

Chapter 6

1 Originally installed in 1776 in the Anatomy Museum of the Institute of Sciences in Palazzo Poggi, after the institute's dissolution by the Napoleonic regime in 1796, the collection of wax anatomies was transferred to the university under the jurisdiction of the Department of Human Anatomy. It was housed in the Academy of Art in the former Jesuit novitiate of St. Ignatius (the current site of the National Pinacoteca). In 1907 it moved to the university's Institute of Human Anatomy in Via Irnerio. In 2000, the collection was restored to its original site in Palazzo Poggi, where it remains today. On this subject, see Armaroli, *Le cere anatomiche bolognesi*, 62.

2 George Gordon Byron, 6th Baron (1788–1824), *"So late into the night": 1816–1817, Letters and Journals*, ed. Leslie A. Marchand (London: William Clowess and Sons, 1976), 5:231–35.

3 Undoubtedly, the most illustrious tourist to visit Morandi's studio was Emperor Joseph II, who, as will be seen in chapter 7, paid unexpected homage to the woman and her collection of anatomical waxes for two hours on Pentecost, 14 May 1769, in her new residence in the palace of Count Girolamo Ranuzzi.

4 Models of the kidneys from this series are extant. Documents suggest that the other models of the male urogenital system may have been destroyed during the mid-nineteenth century by the director of the Institute of Sciences, who condemned the indecency of Lelli's Adam and Eve figures, which he had covered in fig leaves. This document was archived at ASB, Archivio Studio, *Musei e stabilimenti scientifici–Ostetricia* (1804–24), Cart. 462, 468, Titolo IV, but has recently disappeared.

5 Katharine Park, "Dissecting the Female Body: From Women's Secrets to the Secrets of Nature," in *Crossing Boundaries: Attending to Early Modern Women*, ed. Jane Donawerth and Adele Seeff (Newark: University of Delaware Press, 2000), 29.

6 Ludmilla Jordanova, *Sexual Visions: Images of Gender in Science and Medicine between the Eighteenth and Twentieth Centuries* (Madison: University of Wisconsin Press, 1989); Elaine Showalter, "The Woman's Case," in *Sexual Anarchy: Gender and Culture at the fin de Siecle* (New York: Penguin Books, 1990), 127–43; Karen Newman, *Fetal Positions: Individualism, Science, Visuality* (Stanford, CA: Stanford University Press, 1996).

7 On the wax obstetrical models and the school and museum for which they were made, see *Ars obstetricia bononiensis: Catalogo ed inventario del Museo Ostetrico Giovanni Antonio Galli* (Bologna: CLUEB, 1988).

8 Jordanova, *Sexual Visions*, 55.

9 Regnier de Graaf, *On the Human Reproductive Organs* [1668–72], ed. and trans. H. D. Jocelyn and B. P. Setchell (Oxford: Blackwell Scientific Publications, 1972).

10 Showalter, "The Woman's Case," 127–43.

11 Newman, *Fetal Positions*, 33.

12   Ibid., 115.

13   Newman denies the validity of Claudia Pancino's well-documented study of the history of Galli's obstetrical practice and museum, which aimed to diminish infant mortality by educating midwives and surgeons in safe delivery practices.

14   Among other relevant primary materials, there is extensive correspondence by Galli on the subject of the uterine models, as well as contracts, bills of sale, etc., for their creation. Primary documents directly relevant to the commission are, among others, an important series of letters of 27 July 1757–1 May 1758 written by Galli to the Flaminio Scarselli, professor of rhetoric and Ambassador of the Bolognese Senate to Rome, in which Galli seeks to negotiate the terms of the acquisition by the Institute of Sciences of his obstetrical collection and practice. The letters of 20 and 24 August 1757 make explicit reference to the wax modelers who worked with him. BUB, MS 72, Scarselli, *Lettere a Flaminio Scarselli*, 1:98–109. On Pope Benedict XIV's advocacy and the Senate's approval of the establishment of Galli's professorship in obstetrics in the Institute of Sciences, see ASB, Assunteria di Istituto, *Diversorum*, busta 15, no. 34 (1757 e legg. 1800).

15   Newman considers sadomasochistic the representation of isolated, and what she calls "morcellated," components of the body: eyes, legs, ears, hearts, etc. However, as I have discussed elsewhere and also consider here, the anatomical studies by Anna Morandi and Giovanni Manzolini are serial, beginning with the intact component or organ under consideration: the appendage, the sense organs, the cardiovascular or urogenital system, etc., and proceed progressively as in a dissection, with subsequent models demonstrating deeper and more detailed views of the part or system. All the models, from the figure of a torso to a microscopic detail, represent the animate and indeed dynamic body, and serve as synecdoches for the whole living body.

16   Sheriff, *The Exceptional Woman*, 31.

17   Philip James Bailey, *Festus, A Poem* (London: Routledge, 1901), 546.

18   Gianna Pomata, "Donne e Rivoluzione Scientifica: Verso un nuovo bilancio," in *Corpi e storia: Donne e uomini dal mondo antico all'età contemporanea* (Rome: Viella, 2002), 165–91; and "Perché l'uomo è un mammifero: Crisi del paradigma maschile nella medicina di età moderna," in *Genere e mascolinità: Uno sguardo storico* (Rome: Bulzoni, 2000), 133–52.

19   Katharine Park and Robert Nye, "Destiny Is Anatomy," review of *Making Sex: Body and Gender from the Greeks to Freud*, by Thomas Laqueur, *New Republic*, February 18, 1991, pp. 53–57; Park, *Secrets of Women*, 186–87; and Laqueur, *Making Sex*.

20   Pomata, "Donne e Rivoluzione," 167.

21   Ibid., 172.

22   Ibid., 174. In her positive valuation of Antoni van Leeuwenhoek's discovery of seminal animalcoli for the advancement of conceptualizations of sexual difference, Pomata does not discuss the ways in which this "discovery" exceeds even Aristotelian theories in rendering woman irrelevant to human generation except as nutritive matter. For van Leeuwenhoek, within the spermatozoon was preformed man possessed of a living soul, while ovaries were ornamental and the uterus and female genitalia served only to receive and nourish male seed. On the subject, see E. G. Ruestow, "Images and Ideas: Leeuwenhoek's Perception of the Spermatozoa," *Journal of the History of Biology* 16 (1983): 185–224.

23 Pomata, "Donne e Rivoluzione," 173.

24 Ibid., 180.

25 Luciano Guerci, *La discussione sulla donna nell'Italia del Settecento* (Turin: Tirrenia Stampatori, 1987), 130.

26 Early modern and contemporary debates over the emergence of sexually differentiated anatomy are strikingly represented in the arguments and counterarguments expressed by Stolberg, "A Woman Down to Her Bones"; Lacqueur, "Sex in the Flesh"; and Schiebinger, "Skelettestreit."

27 On Galli's obstetrical practice and museum, see *Ars obstetricia bononiensis*.

28 In the commission officially contracted with Giovanni Manzolini but on which Morandi presumably also worked, Galli stipulated the parts to be demonstrated and required that these be modeled on the illustrations in obstetrical treatises by Deventer and Mauriceau. See the following three articles: Bortolloti, "Il maestro alla lavagna"; Pancino, "L'ostetricia del Settecento"; and Lanzarini, "Un museo per la didattica e la sanità ostetrica."

29 For an adroit summary overview of ancient to early modern theories of the uterus's structure and function, see Schiebinger, "Skeletons in the Closet," 42–82.

30 Instrumental for my summary and analysis is the excellent study of the tract by Cavazza, "Women's Dialectics, or the Thinking Uterus." In 1750, Maria Gaetana Agnesi was offered the chair of public lecturer in mathematics at the University of Bologna by Pope Benedict XIV. Although she declined the offer, she attained certain celebrity in the city as a result. On this offer and the pope's correspondence with Agnesi, see Mazzotti, "Maria Gaetana Agnesi,", p. 680, n. 52; and Cavazza, "Women's Dialectics, or the Thinking Uterus," 6–9.

31 Zecchini, *Dì geniali*, 102. Zecchini rejects the brain as the organ of intellection and theorizes that the mind occupies the entire nervous system and can thus "receive impressions from any innervated part of the body to form its ideas" (p. 92).

32 On theories of the wandering uterus, see Lana Thompson, *The Wandering Womb: A Cultural History of Outrageous Beliefs about Women* (Amherst, MA: Prometheus Books, 1999).

33 Zecchini, *Dì geniali*, 114–15. It is noteworthy that this tract elicited responses both within and outside Bologna, including a stinging rebuttal by Giacomo Casanova. See Giacomo Casanova, *Lana Caprina. Epistola di un licantropo indiritta a S. A. la Signora Principessa J. L. N. P. C. Ultima edizione, in nessun luogo* [Venice], *l'anno 100070072* [1772]. Another noted rebuttal was written by Germano Azzoguidi, *Lettres de Madame Cunégonde éscrites de B [...] à Madame Paquette à l'occasion d'un livre qui a pour titre "Dì geniali della dialettica delle donne ridotta al suo vero principio" imprimé a Boulogne chez les frères Taruffi, Paris, 1789*. The publication place and date are false, as this was published in Bologna in 1772.

34 Jordanova, *Sexual Visions*.

35 On the significance of the female body in the anatomical dissection scene, see Park, "Dissecting the Female Body," 29–47; Sawday, *The Body Emblazoned*; and Carlino, *Books of the Body*, esp. 8–68.

36 Sawday, *The Body Emblazoned*, 222.

37 Park, *Secrets of Women*, 93.

38 Giampietro Zanotti, *Storia dell'Accademia Clementina di Bologna, aggregata all'Istituto delle Scienze e delle Arti*, (Bologna: Lelio della Volpe 1739), 2:101.

39 Each of these authors contributed influential tracts claiming to demonstrate scientifically women's essential inferiority: Conti, "Lettera dell'Abate Conti P. V." Galiani wrote a semiserious, hyperbolically misogynist tract in response to Antoine Leonard Thomas's *Essai sur le femme*: "Croquis d'un dialogue sur les femmes" [1772]. Benvenuto Robbio's *Disgrazie di donna Urania* (Parma: Bodniani, 1793) savaged all women with pretenses to learning on the basis of their ontological inferiority. On these and other eighteenth-century tracts focused on female inferiority, see Messbarger, *The Century of Women*, 49–68.

40 Park, "Dissecting the Female Body," 38–9.

41 Andreas Vesalius, *De humani corporis fabrica*, forthcoming online translation by Daniel Garrison and Malcolm Hast, at Northwestern University. Web site http://vesalius.northwestern.edu/index.html.

42 Vesalius, trans. Garrison, 175.

43 Ibid., 78.

44 Although Morandi does not cite de Graaf directly, she clearly knew his theories from extensive references to them in the works of Marcello Malpighi and Giovanni Battista Morgagni, whom she knew well and does cite explicitly.

45 Pomata, "Donne e Rivoluzione," 165–91; and "Perché l'uomo è un mammifero," 133–52.

46 "The common function of the female 'testicles' is to generate eggs, foster them and bring them to maturity. Thus, in women, they perform the same task as do the ovaries of birds. Hence they should be called women's 'ovaries' rather than 'testicles,' especially as they bear no similarity either in shape or content to the male testicles properly so-called." De Graaf, *On the Human Reproductive Organs*, 135.

47 On de Graaf, Malpighi, and other sixteenth- to eighteenth-century theorists who expounded on the female reproductive organs, especially the ovaries, see, R. H. F. Hunter, *Physiology of the Graafian Follicle and Ovulation* (Cambridge: Cambridge University Press, 2003), 3–16.

48 G. B. Morgagni, *La generazione nel concetto di G. B. Morgagni*, ed. and trans. Renato Mighetti and Tommaso Mola (Rome: Istituto di Storia della Medicina dell'Università di Roma, 1963).

49 Antonio Vallisneri, *Istoria della generazione dell'uomo, e degli animali, se sia da' vermicelli spermatici o dalle uova* (Venice: Gio. Gabbriel Herta, 1721), 2.

50 Ibid., 96.

51 Ibid., 3.

52 Ibid., 10.

53 Ibid., 83. Vallisneri is citing and seconding the words of Lorenzo Patarol.

54 Ibid., 90–91.

55 Ibid., 35–36.

56 On Vallisneri's ideas on generation, preexistence, and mechanism, see Jacques Roger, *The Life Sciences in Eighteenth Century French Thought*, ed. Keith R. Benson and trans. Robert Ellrich (Stanford, CA: Stanford University Press, 1997).

57 De Graaf, *On the Human Reproductive Organs*, 100.

58 Maryanne Cline Horowitz, "Aristotle and Woman," *Journal of the History of Biology* 9, no. 2 (1976): 183–213.

59 Morgagni, *La generazione nel concetto di G. B. Morgagni*, 29–30.

60 Catherine Wilson, *The Invisible World: Early Modern Philosophy and the Invention of the Microscope* (Princeton: Princeton University Press, 1995), 122.

61 Marcello Malpighi, *consulto*, 15 April 1693, cited in Howard Adelmann, ed., *Marcello Malpighi and the Evolution of Embryology* (Ithaca, NY: Cornell University Press, 1966), 2:862–64, n. 1. Clara Pinto-Correia provides a brilliant analysis of early modern generative theories and the rich intertextual context in which they developed in *The Ovary of Eve: Egg and Sperm and Preformation* (Chicago: University of Chicago Press, 1997).

62 On the debate and women's responses to it, see Messbarger, *The Century of Women*, chap. 1, "The Debate," 21–48.

63 Vallisneri, *Istoria*, 39.

64 BUB, MS 2193, Morandi, *Anatomical Notebook*, fol. 96r.

65 Ibid., fol. 96v.

66 John C. Weber, *Shearer's Manual of Human Dissection*, 8th ed. (New York: McGraw-Hill, 1990), 127–28.

67 As defined in the *OED* online, *emulgent* signifies:

> A. *adj.* That 'milks out'; *esp.* 'applied to the vessels of the kidneys, which are supposed to strain or milk the serum through the kidneys' (*Syd. Soc. Lex.*). 1578 BANISTER *Hist. Man v. 82* The Emulgent veynes. 1621 BURTON *Anat. Mel. I. i. II. ii*, The branches of the Caua are . . . inward seminall or emulgent. 1670 *Phil. Trans. V. 2081* Passages, by which the Chyle may come into the Emulgent . . . Vessels. 1675 EVELYN *Terra (1776) 23* The Fibres . . . are as it were the Emulgent veins.

> In volume 2 of his *Dictionary of the English Language* (London: Longman, Hurst, Rees, and Orme, 1805), Samuel Johnson provides the most complete eighteenth-century definition of emulgent vessels: "the two large arteries and veins that arise the former from the descending trunk of the aorta, or great artery; the latter from the vena cava. They are both inserted into the kidneys, the emulgent arteries carrying blood with the serum to them, and the emulgent veins bringing it back again, after the serum has been separated therefrom by the kidneys."

68 Eighteenth-century Italian anatomists commonly made reference to "adipose veins and arteries," as seen in Giovanni (Johann) Vesling's *Tavole anatomiche* (Padova: G. B. Conzatti, 1802), a text in Morandi's own archive that she consulted for this series.

69 According to de Graaf, "In men, the testicles normally hang forward outside the abdomen, beneath the lower belly at the root of the penis in the scrotum, which envelops them, so that because of the length of the duct the ingredients of the semen have to linger about a longer time and thus receive a better preparation." *On the Human Reproductive Organs*, 12.

70 BUB, MS 2193, Morandi, *Anatomical Notebook*, fols. 96v–97r. I wish to thank Adam Kibel, MD, associate professor of urology at Washington University, for his generous help with the translation and interpretation of this text.

71 ASB, Archivio Ranuzzi, *Istrumenti scritture*, Libro 124, no. 21, "Inventario dei libri venduti dalla Signora Manzolini al Sig.r Sen.re Girolamo Ranuzzi, con il suo Armario."

72   Andrea Carlino, "Vesalio e la cultura visiva delle anatomie a stampa del Rinascimento," in Olmi, *Rappresentare il corpo*, 75–91.

73   On the Albertian theories on the place of anatomy in art, see Paola Salvi, "Da Leonardo alle accademie: Procedimenti e metodi anatomici degli artisti," in Olmi, *Rappresentare il corpo*, esp. 51–55.

74   Images of the male reproductive system in Johann Vesling and especially Morgagni, whose works Morandi possessed, may also have influenced her representations. By contrast, despite her familiarity with William Cowper's atlas, the graphic depictions of the dead and putrefying male and female reproductive bodies he includes from Govert Bidloo's atlas represent a point of view antithetical to Morandi's always vital body images.

75   Certainly as part of the Bolognese community of medical professors and practitioners, she would have had some knowledge of the study and preservation of bodies with wax by, among others, Jan Swammerdam and his famous student Frederick Ruysch. On these methods, see Harold J. Cook, *Matters of Exchange: Commerce, Medicine and Science in the Dutch Golden Age* (New Haven, CT: Yale University Press, 2007), esp. 267–303.

76   BUB, MS 2193, Morandi, *Anatomical Notebook*, fol. 99r–v.

77   Ibid., fol. 100v. Galen states:

> The so-called neck of the bladder is muscular and the lower end of the rectum is held shut by circular muscles surrounding it [sphincter ani, externus and internus]. This is the reason, I suppose, why some have called it the sphincter. For all the muscles, being instruments of voluntary motion, do not allow the residues to be evacuated except at the command of reason, and here at the two outlets for the residues is the only instance in this whole long course of the physical (natural) instruments [alimentary tract and urinary organs] where there is an instrument of the psychic soul. If in some individuals these muscles are relaxed or impaired in any other way ever so slightly the residues flow out involuntarily and inopportunely, showing clearly how shameful and gross would be our life if from the beginning Nature had not planned something better.

Claudius Galen, *On the Usefulness of the Parts of the Body*, trans. Margaret Tallmadge May (Ithaca, NY: Cornell University Press, 1968), 241. In chapter 49 of book 2 of the *Fabrica* (page 326, numbered 226), Vesalius identifies, in his figure 1, "Muscle [m. sphincter urethrae externus] of the neck of the bladder, preventing urine from flowing down against our will" (trans. Daniel Garrison, 532).

78   Siraisi, *Medicine and the Italian Universities*, 254.

79   For an astute analysis of ancient and early modern teleological anatomy, see ibid., 253–86.

80   Ibid., 266–69.

81   Ibid., 266–75.

82   De Graaf, *On the Human Reproductive Organs*, 14.

83   Morgagni, *La generazione nel concetto di G. B. Morgagni*, 200.

84   Ibid., 208.

85   Marcello Malpighi, "On the Formation of the Chick in the Egg," appendix to *Repeated and Additional Observations on the Incubated Egg*; in Adelmann, *Marcello Malpighi*, 2:984–85.

86 Nancy Siraisi's interpretation of the teleology at work in Galen and Vesalius was instrumental for my analysis of Morandi's own teleological concerns: *Medicine and the Italian Universities*, 254–85.

87 Messbarger, "Waxing Poetic," 80–92.

88 It is important to note that Morandi was by no means the first to make these observations about the positioning of the testicles. BUB, MS 2193, Morandi, *Anatomical Notebook*, fol. 97r.

89 BUB, MS 2193, Morandi, *Anatomical Notebook*, fol. 98v. Indeed she was quite accurate in her description of this part of the male anatomy.

90 As noted previously, wax injections were not an uncommon practice among anatomists by this point in time.

91 Ibid., fol. 117r.

92 Ibid., fol. 117r–v. Here again, Morandi misunderstood the role of the seminal vesicles, through which semen do not pass, but from which semen receive a necessary nutrient.

93 Galvani, "De Manzoliniana," 100.

94 Semen resides in the epididymis before ejaculation and bypasses the seminal vesicles. My thanks to Dr. Adam Kibel for his explanation of this process.

95 BUB, MS 2193, Morandi, *Anatomical Notebook*, fol. 105r. *Balsamic* connoted healing, soothing, and medicinal properties and was used occasionally in anatomical texts to describe semen. For example, in *Istoria*, his tract of 1721 defending spermist arguments for preformation, Vallisneri states that semen "è piuttosto una dolcissima, balsamica, spiritosa sostanza," and later calls it, with Hippocrates, "balsamo-oleoso-volatile" (59, 90).

96 The Italian translation of Jacques Bénigne Winslow's book, *An Anatomical Exposition of the Structure of the Human Body*, trans. G. Douglas (London: N. Prevost, 1733), was in Morandi's archive. On page 549 regarding the seminal vesicles, Winslow states, "Their surface is villous and glandular, and continually furnishes a particular fluid, which exalts and refines, and perfects the semen, which they receive from the vasa deferentia."

97 BUB, MS 2193, Morandi, *Anatomical Notebook*, fol. 117v.

98 Ibid., fols. 116r–119r. Sperm travels from the testicles through the epididymis, through the vas deferens and ejaculatory ducts to the prostate and out through the urethra. Along this path, the sperm collects secretions from the seminal vesicles and the prostate that provide necessary nutrients and volume to sperm. My thanks again to Dr. Adam Kibel for explaining this process.

99 On medieval accounts of this process, see Danielle Jacquart and Claude Thomasset, *Sexuality and Medicine in the Middle Ages* (Princeton, NJ: Princeton University Press, 1985), 52–70.

100 In his intricately detailed analysis, de Graaf (*On the Human Reproductive Organs*, 32, 113) avers that the spermatic arteries carry blood to the testicles for the nutrition and generation of semen, and that the nerves that reach the substance of the testicles "carry the animal spirits which are indispensable for generation." Semen, he maintains, is confected from these and other materials in the testicles. He further states that

the ingredients of the semen are propelled through the very long ducts of the testicles, the semen is elaborated in their cavities in such a way that what was watery and ash-like in the testicles becomes milky and thick in the epididymides. This elaboration of the ingredients of the semen occurs . . . through the watery parts being separated from them and through more spirits being added to the mixture. These cause the ingredients to foam more as they pass through the tubules and bestow final perfection upon them. . . . [Thus,] only one kind of semen is generated and this in the testicles alone. This semen is preserved in the vesicles and at an opportune moment is propelled into the urethra, enveloped there with the fluid of the prostate gland and ejaculated further into the uterus.

Morgagni held that "the testicles produce sperm" and that the seminal vesicles, the prostate, Cowper's glands, etc., produce "materie" that are useful for said sperm: *La generazione nel concetto di G. B. Morgagni*, 141–49.

101   De Graaf, *On the Human Reproductive Organs*, 32:

We think that the nerves run out through the very thin membranes which hold steady the spermatic vessels in their course inside the testicles, or perhaps even make up these membranes and through them deposit the animal spirits in the seminiferous tubules. The animal spirits, which have joined with the more subtle part of the blood, pass out simultaneously through the ducts of the seminiferous tubules to the epididymides, the vasa deferentia and thence the penis into the place ordained by Almighty God for the reception of the semen. . . . These cause the ingredients to foam more as they pass through the tubules and bestow final perfection upon them, a process we believe Hippocrates to have understood similarly to ourselves where he says that foam is the "essence of semen."

102   Absent from her account are any theories remotely similar to Galen's, that although "all the parts that men have, women have too . . . of course the female must have smaller, less perfect testes, and the semen generated in them must be scantier, colder," and thus "incapable of generating an animal" (Galen, *On the Usefulness of the Parts*, 2:630, 631). Like Galen, Vesalius viewed female semen as inferior to the male although necessary for reproduction.

103   Jacquart and Thomasset, Sexuality and Medicine in the Middle Ages, 60.

Chapter 7

1   BCAB, MS B 120, Marcello Oretti, *Lettere di diversi al S.r Oretti*, fol. 183. Among Oretti's correspondence in this same section are letters Morandi apparently wrote for the Senate to plead her case directly. One of these states: "Finding herself in extreme need of aid because of the long illness that she has suffered for months as well as other costly circumstances, Anna Manzolini, salaried public lecturer of this celebrated University of Bologna, therefore ardently implores your most illustrious lords to deign to agree to a supplement of two hundred lire so that she might live in such a way as to permit her to toil even more, as has been and remains her desire, both for the public utility of her art and for the Patria" (fol. 190).

2 Pope Benedict first intervened on Morandi's behalf to advocate the original annual 300-*lire* honorarium she had received in 1755 "to provide assistance with her domestic circumstances so that she could continue with greater courage her studious labors and so that she would decline the invitations being sent to her by other Countries." Catherine the Great's invitation was among these. ASB, Gabella Grossa, *Atti delle Congregazioni I/42*, 15 December 1755, fol. 112.

3 Crespi writes of the blank check Morandi received from Milan to move her practice there. Fantuzzi says that she was invited to practice there. Crespi, *Felsina pittrice*, 311; Fantuzzi, *Notizie degli scrittori bolognesi*, 6:115.

4 BCAB, MS B 133, vol. 11, Oretti, *Notizie de' professori*, fol. 228.

5 On Ranuzzi see Anna Rosa Bambi, "Il Conte Girolamo Ranuzzi: Un eclettico Bolognese del '700,'" *Il Carrobbio* 24 (1998): 137–56.

6 ASB, Archivio Ranuzzi, *Istrumenti scritture*, Libro 124, no. 6.

7 Ibid.

8 Ibid., Libro 125, no. 23. On Isolani, see C. Barbieri, *Memorie della vita e delle virtù del servo di Dio Ercole M. G. Isolani, prete della Congregazione dell'Oratoria di Bologna* (Venezia, 1711). Although the existence of the Isolani bust is currently uncertain, it was still extant and listed as part of the exhibition of eighteenth-century Bolognese art in 1935: *Mostra del Settecento Bolognese, Catalogo* (Bologna: Mareggiani, 1935), 146.

9 ASB, *Diversorum*, busta 10, no. 5, "Notizie attinenti allo Studio Manzolini. Acquisto di esso fatto da SS.ri dell'Istituto, 1775, 1776, e 1777."

10 Archivio di San Procolo, Amministrazione Parrocchiale, *Stato delle anime*; among the documents in the archive relating to the Ranuzzi household is the parish census information for the years 1769–74. In these are listed as occupants of the Ranuzzi household Anna Morandi Manzolini, her son Carlo, whose age in 1769 is given to be twenty-one, and Morandi's maid, Lucia Grifoni, age thirty-five.

11 My abundant thanks to Prof. Sergei Karp for generous help with this information.

12 ASB, Archivio Ranuzzi, *Lettere di vari sovrani* (T. D., p. 3, Sc. 17). It should be noted that these letters have not been formally catalogued. I thank Anna Bambi and Romolo Dodi for indicating these to me.

13 Claiming that these were his property and that their dispersal to another owner would diminish the value of his complete collection, after Morandi's death he thus sought to block their shipment to Breslau. ASB, Archivio Ranuzzi, *Carte di Famiglia*, busta 137, "Lettere, Pro-memoria a Monsignore Ranuzzi, La Principessa Jablonowska."

14 Although Gabriella Berti Logan makes a case for the strategic move of Morandi to the Ranuzzi palace by civic leaders in anticipation of the emperor's visit, accounts of the royal visit clearly indicate that Joseph II had given no advance notice of his trip and was traveling under the pseudonym Count of Wolkestein. Records of the Institute of Sciences state: "His Majesty, the Emperor Joseph II, who passed through Bologna without anyone's knowledge, and who having gone to Rome perfectly incognito, returned to Bologna arriving the evening of the 9th of May and lodged at the Locanda del Pellegrino . . . He then went to Parma for two days and returned to Bologna Saturday, the 13th, again lodging at the Locanda del Pellegrino." ASB, Assunteria di Istituto, *Diversorum*, busta 18, no. 16, "Visita all'Istituto dell'Imperatore Giuseppe II" (1769).

15 The inscription on the medallion is described by Marcello Oretti in his biographical sketch "Anna Morandi." BCAB, MS B 133, vol. 11, *Notizie de' professori*, fol. 228.

16 On Fontana and the Specola wax anatomies, see Renato Mazzolini, "Plastic Anatomies and Artificial Dissections," in *Models: The Third Dimension of Science*, ed. Soraya de Chadarevian and Nick Hopwood (Stanford, CA: Stanford University Press, 2004), 43–70; and Maerker, "Model Experts."

17 ASB, Archivio Ranuzzi, *Istrumenti scritture*, Libro 124, no. 15.

18 Ranuzzi had doubtless sought and failed to sell the bust to her five years before, when it was first cast. It is unclear if the empress had subsequently agreed to purchase the bust, perhaps for a more reasonable price, or if Ranuzzi was sending it as a gift.

19 Ranuzzi's autograph letter was found by the scholar Sergei Karp in the recently accessible diplomatic archives (Arkhiv vneshnej politiki Rissijskoj Imperii, AV-PRI, Moscow). I am deeply indebted to Prof. Karp for sharing this letter with me. For the translation of Ranuzzi's difficult and convoluted letter, I am very grateful to Pascal Ifri and Lynne Breakstone.

20 J. Grot, *С oр Pyccко o пeрaмoрcкo иcmoрu ecкo o o ecm a* (St. Petersburg, 1878), 52.

21 On the history of the bust at Peterhof and for one of the remaining pictures of it, see Villi Umangulov, "Sculpture of Anna Manzolini Work by Nollekens in Petergof," in *Study Group on Eighteenth-Century Russia, Newsletter* 29 (September 2001): 55–67. My sincere thanks to Elena Mozgovaja for information about this article, and to Marina Isupov for translating it into English.

22 Pietro Metastasio, *Tutte le opere di Pietro Metastasio* (Milano: Mondadori, 1951–65), letter 1911, p. 61. He declined to help her.

23 ASB, Ufficio del Registro, Copie degli atti, Libro 921, *Testamento Anna Morandi Manzolini* (9 July 1774).

24 Translation from the Latin by George Pepe.

25 Archive of the Church of San Procolo, Cartone 44, fasc. 27. Anno 1872–73, *Apertura della tomba e transporto nel Cimitero Comunale delle ossa della celebre Anna Manzolini*. According to documents of the Administrative Office of the Municipal Cemetery of Bologna, at 7:45 a.m. on 21 March 1873, the remains of Anna Morandi Manzolini were brought to the Certosa Cemetery and placed in the *tombino*, or small tomb, already opened for the purpose, number 971.

26 Bosi, *Archivio patrio*, 245.

27 ASB, Assunteria d'Istituto, *Atti*, vol. 6 (1761–75).

28 ASB, Archivio Ranuzzi, *Istrumenti scritture*, Libro 125, no. 23, 25 June 1776.

29 Galvani, "De Manzoliniana," 94–103.

30 Ibid, 99.

31 Ibid.

32 Aristotle, *On Man in the Universe*, ed. Louise Ropes Loomis (New York: Walter J. Black, 1943), 423. My thanks to Ronald Briggs for suggesting this link.

33 Lucia Dacome discusses the Galvani oration in "Waxworks and the Performance of A," 30; and in "Women, Wax and Anatomy," 527.

34 Galvani, "De Manzoliniana," 99.

35 Ibid., 100.

36 Ibid.

37 On the history of the Specola, there are a number studies of note. Especially useful is the comprehensive study by Simone Contardi, *La casa di Salomone a Firenze:*

L'Imperiale e Reale Museo di Fisica e Storia Naturale (1775–1801) (Florence: Olschki, 2002). See also Didi-Huberman, von Düring, and Poggesi, *Encyclopaedia Anatomica*; and Peter Knoefel, *Felice Fontana, Life and Works* (Trento: Società di Studi Trentini di Scienze Storiche, 1984).

38 Museo della Specola, *Giornali dei modellatori*, 1793, 1796, 1797, 1798. My thanks to Dr. Marta Poggesi for allowing me to study these manuscripts on display in the museum.

39 Maerker, "The Anatomical Models of La Specola," 300–301; and "Model Experts."

40 Fontana's letter to Caldani, 30 October 1791, is reproduced in Renato G. Mazzolini and Giuseppe Ongaro, eds., *Epistolario di Felice Fontana*, vol. 1, *Carteggio con Leopoldo Marc'Antonio Caldani, 1758–1794* (Trento: Società di Studi Trentini di Scienze Storiche, 1980), 318–19.

41 Although there is ample evidence that he held this view, Fontana did engage in most aspects of the model production. Maerker provides an insightful analysis of the superior cognitive authority claimed by natural philosophers over the technical skills of Specola artisans in "Model Experts," 63–115.

Bibliography

Primary Sources

Manuscripts

ARCHIVIO DI STATO, BOLOGNA (ASB)

Anziani consoli: *Insignia degli Anziani del Comune dal 1530–1796*, vol. 13, cc. 94a, 105a.

Archivio Ranuzzi: *Carte di Famiglia, Lettere*, busta 137.

Archivio Ranuzzi: *Istrumenti scritture diverse spettanti alla Nobile Casa Ranuzzi dall'anno 1769 all' anno 1773*, Libro 124, 125.

Archivio Ranuzzi: *Lettere di vari sovrani*.

Archivio Studio: *Musei e stabilimenti scientifici–Ostetricia* (1804–24), Cart. 462, 468, Titolo IV.

Assunteria d'Istituto: *Atti*, vol. 3 (1727–34).

Assunteria d'Istituto: *Atti*, vol. 4 (1734–53).

Assunteria d'Istituto: *Atti*, vol. 5 (1754–60)

Assunteria d'Istituto: *Atti*, vol. 6 (1761–75).

Assunteria d'Istituto: *Diversorum*, busta 10, no. 12, "Parallelo fra l'Università di Bologna e le oltramontane del Co. L. Ferd. Marsili," 1709.

Assunteria d'Istituto: *Diversorum*, busta 10, no. 1, "Proposta per le statue anatomiche," 12 February 1732.

Assunteria d'Istituto: *Diversorum*, busta 10, no. 1, "Ercole Lelli, All'illustrissima ed eccellentissma Assunteria dell'Istituto delle Scienze di Bologna—Proposta per la nuova Camera Anatomica," 5 October 1742.

Assunteria d'Istituto: *Diversorum*, busta 10, no. 2, "Fondazione delle Camere di Anatomia fatte da Benedetto XIV 1742–1747 statue anatomiche formate in cera da Ercole Lelli," 1 December 1742.

Assunteria d'Istituto: *Diversorum*, busta 10, no. 2.2, "Ristretto delle Mercedi e spese da darsi e da farsi per formare le statue di cera per la nuova Stanza di Notomia dell'Istituto."

Assunteria d'Istituto: *Diversorum*, busta 10, no. 5, "Notizie attinenti allo Studio Manzolini. Aquisto di esso fatto da SS.ri dell'Istituto, 1775, 1776, e 1777."

Assunteria d'Istituto: *Diversorum*, busta 10, no. 5, "Compere dal Conte Girolamo Ranuzzi dello Studio e delle preparazioni anatomiche di Anna Manzolini e notizie attinenti."

Assunteria d'Istituto: *Diversorum*, busta 10, "Serie Anna Manzolini," 1771.

Assunteria d'Istituto: *Diversorum*, busta 15, no. 34 (1757 e legg. 1800).

Assunteria d'Istituto: *Diversorum*, busta 18, no. 16, "Visita all'Istituto dell'Imperatore Giuseppe II," 1769.

Assunteria dei Magistrati: *Affari Diversi*, busta 78, no. 3, "Benedictus Papa XIV Motu Proprio," 28 November 1747.

Assunteria di Studio: *Diversorum*, busta 91, "Anatomia Pubblica," 1750–55.

Assunteria di Studio: *Diversorum*, busta 91, "Relazione degli Assunti ed altra circa gli Anatomici gli interesse egli [...] dell'Anatomia," May 1755.

Assunteria di Studio: *Diversorum*, busta 91, "Ceremoniale per le lezioni di anatomia," 1765.

Gabella Grossa: *Libri Segreti*, 1628–1640.

Gabella Grossa: *Atti delle Congregazioni I/42*, 15 December 1755.

Orfanotrofio di S. Maria Maddalena e di S. Bartolomeo uniti: *Statuti e ammissioni*, busta 2, "Campione dei puti," fol. 26.

Ufficio del Registro: Copie degli atti, Libro 921, *Testamento Anna Morandi Manzolini* (9 July 1774).

ARCHIVIO ACCADEMIA BELLE ARTI DI BOLOGNA

Atti dell'Accademia Clementina, vol. 1, fol. 190, 3 December 1755.

Atti dell'Accademia Clementina, vol. 1, fols. 65–66.

ARCHIVIO ARCIVESCOVILE, BOLOGNA (AAB)

Libri battesimali (1690–1742).

Libri mortuorum (1743–1805).

Parocchie di Bologna soppresse, San Nicolò degli Albari, Cart. 35/8, 3.

ARCHIVIO DI SAN PROCOLO

Amministrazione Parrocchiale: *Stato delle anime* (1769–74).

Cartone 44, fasc. 27 (1872–73): *Apertura della tomba e transporto nel Cimitero Comunale delle ossa della celebre Anna Manzolini.*

BIBLIOTECA COMUNALE DELL'ARCHIGINNASIO, BOLOGNA (BCAB)

MS B 95 no. 26, fol. 119, 120. Oretti, Marcello. *Notizie de professori ecc.*, vol. 14, *Vite di pittori scultori e architetti in gran parte scritte da loro medesimi raccolte da Marcello Oretti.*

MS B 110. Oretti, Marcello. *Pitture nelli palazzi del territorio bolgonese* (n.d.).

MS B 120, fols. 182–92. Oretti, Marcello. *Lettere di diversi al S.r Oretti* (n.d.).

MS B 133, vol. 11, fols. 227–30. Oretti, Marcello. *Notizie de professori del disegno cioè pittori, scultori ed architetti bolognesi e de' forestieri di sua scuola raccolte ed in più tomi divise da Marcello Oretti bolognese accademico dell'Istituto delle Scienze di Bologna* (n.d.).

MS B 134, vol. 12, fols. 133–36. Oretti, Marcello. *Notizie de' professori dell'arte del disegno* (n.d.).

MS B 1562. Lelli, Ercole. *Compendio anatomico per uso delle Arti del disegno di Ercole Lelli bolognese pittore, scultore e Disegnatore famosissimo* (n.d.).

MS Fondo Mondini: *Richiesta all'Ospedale di S. Maria della Morte dell'Amm.re C.o Vittori per procurar cadaveri alla Anna Manzolini*, Cart. VIII, n. 3 (1756).

MS Fondo Ospedale: *Giustiziati dall anno 1674–1796*.

MS Collezione degli autografi LXV, 17.584–17.754 numero progressivo; 17.710, 17.712 copia autografa.

Regolamenti per gli esercizi letterarj dell'Accademia de' Gelati Restituiti nella sessione del 31 luglio 1786, p. 6.

BIBLIOTECA UNIVERSITARIA, BOLOGNA (BUB)

MS 2193. *Catalogo delle preparazioni anatomiche in cera formanti il gabinetto anatomico prima della Reggia Università* [Anna Morandi Manzolini, *Anatomical Notebook*] (n.d.).

MS 90, A 10. Letter from Benedict XIV to Ercole Lelli, Rome, 15 January 1751.

MS 3882, caps. LVIII A 5. *Cinque lettere da Roma di Marco Antonio Laurenti a Ercole Lelli* (1747–53).

MS 72. Scarselli, Carteggio. *Lettere a Flaminio Scarselli*, vol. I (1742–60).

MS 243. Beccari, Jacopo Bartolomeo. *Lettere autografe al dott. Flaminio Scarselli* (1740–61).

MS 75, II, cc. 323–38. Marsili, Anton Felice. *Discorso dell'apertura delle due accademie in casa di Mons. Arcidiacono Ant. Felice Marsili, l'una ecclesiastica e l'altra filosofica nel mese di novemb. 1687 e dal medesimo Monsignore derivato*.

MS 239, I Personaggi, fasc. VI, no. 12. *Iscrizioni sopra l'ingresso delle case degli Infrac*.

BIBLIOTECA CIVICA DI GAMBALUNGHIANA, RIMINI

Gambetti, Fondo. *Lettere autografe al dott. Giovanni Bianchi*, fasc. Manzolini Morandi, Anna e Giovanni Manzolini. 15 April 1755 and 24 May 1755.

ARCHIVIO DI STATO, FIRENZE

Accademia del disegno. *Primi compagni di pittori*.

MUSEO DELLA SPECOLA

Giornali dei modellatori, 1793, 1796, 1797, 1798.

Published Works

Algarotti, Francesco. *An Essay on Painting Written in Italian by Count Algarotti*. London: L. Davis and C. Reymers, 1763.

———. *Il Newtonianismo per le dame ovvero dialoghi sopra la luce i colori e l'attrazione*. Naples: Giambattista Pasquali, 1739.

Alighieri, Dante. *Paradiso*, canto XIII. Translated by Henry Wadsworth Longfellow. Boston: Houghton Mifflin, 1886.

Angelelli, Giuseppe. *Notizie dell'origine e progressi dell'Istituto delle Scienze di Bologna e sue accademie*. Bologna: Istituto delle Scienze, 1780.

Anonymous. *La difesa delle donna*. Milan, 1767.

Aristotle. *On Man in the Universe*. Edited by Louise Ropes Loomis. New York: Walter J. Black, 1943.

Astruc, Jean. *Traité des maladies des femmes*. 6 vols. Paris, 1761–65.

I. Romanelli. In *Atti del XXVII Congresso Nazionale di Storia della Medicina*. Caserta-Capua-Salerno, Italy, September 1975.

Mauriceau, François. *Description anatomique des parties de la femme, qui servent à la generation; avec un traité des monstres, de leur causes, de leur nature, & de leur differences: Et une description anatomique*. A Leide, Chez la Veve de Bastiaan Schouten, 1708.

―――. *Traité des maladies des femmes grosses, et de celles qui sont nouvellement accouchées*. Paris, 1668.

Metastasio, Pietro. *Tutte le opere di Pietro Metastasio*. Milan: Mondadori, 1951–65.

Moore, John. *A View of the Society and Manners in Italy*. London: Strahan and Cadell, 1795.

Morgagni, Giovanni Battista. *Carteggio inedito di G. B. Morgagni con Giovanni Bianchi*. Edited by Guglielmo Bilancioni. Bari: Società Tipografica Editrice Barese, 1914.

―――. *La generazione nel concetto di G. B. Morgagni*. Edited and translated by Renato Mighetti and Tommaso Mola. Rome: Istituto di Storia della Medicina dell'Università di Roma, 1963.

Morgan, John. *Journal of Dr. John Morgan of Philadelphia from the City of Rome to the City of London 1764*. Philadelphia: J. B. Lippincott, 1907.

Newton, Isaac. *Opticks*. New York: Courier Dover, 1979.

Pisarri, Carlo. *Dialoghi tra Claro e Sarpiri per istruire chi desidera d'essere un eccellente pittore figurista*. Bologna: F. Pisarri, 1778.

Rivinus, Johannes Augustus Quirinus. *De auditus vitiis*. Leipzig, 1717.

Robbio, Benvenuto. *Disgrazie di Donna Urania*. Parma: Bodniani, 1793.

Roederer, Johann Georg. *Elementa artis obstetriciae in usum auditorum denuo edidit, nec non praefatione et annotationibus instruxit H. A. Wrisberg*. Göttingen: Albizziniana, 1759.

Shakespeare, William. *King Richard II*. Edited by Andrew Gurr. Cambridge: Cambridge University Press, 2003.

―――. *Romeo and Juliet*. Edited by Howard Furness. Philadelphia: Lippincott, 1899.

Vallisneri, Antonio. *Istoria della generazione dell'uomo, e degli animali, se sia da' vermicelli spermatici o dalle uova*. Venice: Gio. Gabbriel Herta, 1721.

Valsalva, Antonio Maria. *De aure humana tractatus*. Bologna: Ex officina Guillielmi vande Water, 1707.

Vasari, Giorgio. *Vite de' più eccellenti architetti, pittori, et scultori italiani*. Firenze: G. C. Sansoni, 1906.

Vesalius, Andreas. *De humani corporis fabrica*. Translated by Daniel Garrison and Malcolm Hast. Evanston, IL. Northwestern University Press, in progress. http://vesalius.northwestern.edu/index.html.

Vesling, Giovanni. *Tavole anatomiche*. Padova: G. B. Conzatti, 1802.

Vocabulario della Crusca. Firenze: D. M. Manni, 1733–38.

Volkmann, Johann J. *Historisch-kritische Nachtrichten von Italien, welche eine Beschreibung dieses Landes. Der Sitten, Reigierungsform, Handlung, des Zustandes des Wissenschaften und insonderheit der Kunst enthalten*. 3 vols. Leipzig: Caspar Fritsch, 1777.

Volpi, Giovanni A. "Che non debbono ammesttersi le donne allo Studio delle Scienze, e delle belle arti. Discorso accademico del Signor Gio. Antonio Volpi Pubblico Proffesore di Filosofia nello Studio di Padova, da lui recitato in Padova nell'Accademia de' Ricovrati il di 16 giugno 1723." In *Discorsi accademici di vari autori viventi intorno agli studi delle donne*, edited by Giovanni Antonio Volpi, 27–45. Padua: Giovanni Manfrè, 1729.

Weitbrecht, Josias. *Syndesmolgia sive historia ligamentorum corporis humani quam secundum observations anatomicas concinnavit et figures ad objecta recentia admubratis illustravit*. St. Petersburg: Typographia academiae scientiarum, 1742.

Winslow, Jacques Bénigne. *An Anatomical Exposition of the Structure of the Human Body*. Translated by G. Douglas. London: N. Prevost, 1733.

Zanotti, Francesco Maria. *De Bononiensi Scientiarum et Artium Instituto atque academia commentarii*. 7 vols. Bologna: Lelio della Volpe, 1731–91.

Zanotti, Giampietro. "Al Sig. Ercole Lelli da Giampietro Zanotti, in Giovanni Gaetano Bottari" (1757). In *Raccolta di lettere sulla pittura, scultura ed architettura scritte da' più celebri personaggi dei secoli XV, XVI, e XVII, pubblicata da M. Gio. Bottari, e continuata fino ai nostri giorni da Stefano Ticozzi*, vol. 2. Milan: G. Silvestri, 1822–25.

———. *Storia dell'Accademia Clementina di Bologna, aggregata all'Istituto delle Scienze e delle Arti*. 2 vols. Bologna: Lelio della Volpe, 1739.

Zecchini, Petronio Ignazio. *Dì geniali: Della dialettica delle donne ridotta al suo vero principio*. Bologna: A. S. Tommaso d'Aquino, 1771.

Secondary Sources

Adelmann, Howard, ed. *Marcello Malpighi and the Evolution of Embryology*. Ithaca, NY: Cornell University Press, 1966.

———. *Discorso del Vincitore del Premio Internazionale Galileo Galilei dei Rotary Italiani, 1972, Prof. Howard B. Adelmann*. Pisa, 1972. http://www3.humnet.unipi.it/galileo/fondazione/Vincitori%20Premio%20Galilei/Howard_Adelmann.htm.

Alma mater studiorum: La presenza femminile dal XVIII a XX secolo; Ricerche sul rapporto Donna/Cultura Universitaria nell'Ateneo Bolognese. Bologna: CLUEB, 1988.

Ameisenowa, Zofia. *The Problem of the Écorchés and the Three Anatomical Models in the Jagiellonian Library*. Translated by Andrezej Potocki. Warsaw: Zaklad Narodowy Imtenja Ossolinskich, 1963.

Angelini, Annarita, ed. *L'Istituto delle Scienze e l'Accademia*. Vol. 3 of *Anatomie accademiche*. Bologna: Il Mulino, 1993.

Armaroli, Maurizio, ed. *Le cere anatomiche bolognesi del Settecento*. Bologna: Università degli Studi di Bologna, 1981. An exhibition catalogue.

Ars obstetricia bononiensis: Catalogo ed inventario del Museo Ostetrico Giovan Antonio Galli. Bologna: CLUEB, 1988.

Baccilieri, Adriano. "La storia dell'edificio: Dall'Ospedale di S. Maria della Morte al Museo Civico." In *Dalla Stanza della Antichità al Museo Civico*, edited by Cristiana Morigi Govi and Giuseppe Sassatelli, 101–14. Bologna: Grafis Edizioni, 1984.

Baldacci, Sandra, and Valerio Bartoloni. *Giovanni Lami e il Valdarno inferiore*, edited by Valerio Bartoloni. Pisa: Pacini Editore, 1997.

Bagliani, Agostino Paravicini. "Storia della scienza e storia della mentalità: Ruggero Bacone, Bonifacio VIII e la teologia della prolungatio vitai." In *Aspetti della letteratura latina nel secolo XIII: Atti del primo convegno internazionale di studi dell'Associazione per il medioevo e l'umanesimo latini (AMUL)* 15 (1986): 243–80.

Bambi, Anna Rosa. "Il Conte Girolamo Ranuzzi: Un eclettico bolognese del '700.'" *Il Carrobbio* 24 (1998): 137–56.

Barker-Benfield, G. J. *The Culture of Sensibility: Sex and Society in Eighteenth-Century Britain*. Chicago: University of Chicago Press, 1992.

Belloni, Luigi. "Suono e orecchio dal Galilei al Valsalva: Nel Terzo Centenario della nascita di A. M. Valsalva." *Simposi Clinici CIBA* 3, no. 3 (1966): 33–42.

Bernabeo, Raffaele A. "Ercole Lelli (Bologna 1702–1766)." In Armaroli, *Le cere anatomiche bolognesi del Settecento*, 30–32.

———. "La libreria scientifica di Anna Morandi Manzolini." In Armaroli, *Le cere anatomiche bolognesi del Settecento*, 36–39.

Bernabeo, R. A., and I. Romanelli. "Considerazioni di Giovanni Manzolini (1700–1755) sull'anatomia dell'orecchio in condizioni normali e patologiche." In *Atti del XXVII Congresso Nazionale di Storia della Medicina* (Caserta-Capua-Salerno, Italy, 1975), 1–8.

Bortolloti, Marco. "Il maestro alla lavagna: Il museo del Galli dall'inventario al catalogo." In *Ars obstetricia bononiensis*, 14–23.

Boschloo, Anton W. A. *L'Accademia Clementina e la preoccupazione del passato*. Bologna: Nuova Alfa Editoriale, 1989.

Brown, Elizabeth A. "Authority, the Family and the Dead in Late Medieval France." *French Historical Studies* 16, no. 4 (Fall 1990): 803–32.

Calvino, Italo. *Six Memos for the Next Millennium*. Cambridge, MA: Harvard University Press, 1988.

Carlino, Andrea. *Books of the Body: Anatomical Ritual and Renaissance Learning*. Translated by John and Anne Tedeschi. Chicago: University of Chicago Press, 1999.

———. "Vesalio e la cultura visiva delle anatomie a stampa del Rinascimento." In Olmi, *Rappresentare il corpo*, 75–91.

Cavazza, Marta. "Between Modesty and Spectacle: Women and Science in Eighteenth-Century Italy." In *Italy's Eighteenth Century: Gender and Culture in the Age of the Grand Tour*, edited by Paula Findlen, Wendy Wassing Roworth, and Catherine Sama, 275–302. Stanford, CA: Stanford University Press, 2009.

———. "Dottrici e lettrici dell'Università di Bologna nel Settecento." *Annali di storia delle università italiane* 1 (1997): 109–26.

———. "Laura Bassi." Bologna Science Classics Online. www.cis.unibo.it/cis13b/bsco3/bassi/bassinotbyed/bassinotbyed.pdf.

———. "La recezione della teoria halleriana dell'irritabilità nell'Accademia delle Scienze di Bologna." *Istituto e Museo di Storia della Scienza Nuncius: Journal of the History of Science* 12, no. 2 (1997): 359–77.

———. *Settecento inquieto: Alle origini dell'Istituto delle Scienze di Bologna*. Bologna: Il Mulino, 1990.

———. "Women's Dialectics, or the Thinking Uterus: An Eighteenth-Century Controversy on Gender and Education." In *The Faces of Nature in Enlightenment Europe*, edited by Lorraine Daston and Gianna Pomata, 237–57. Berlin: BWV–Berliner Wissenschafts-Verlag, 2003.

Contardi, Simone. *La casa di Salomone a Firenze: L'Imperiale e Reale Museo di Fisica e Storia Naturale (1775–1801)*. Florence: Olschki, 2002.

Cook, Harold J. *Matters of Exchange: Commerce, Medicine and Science in the Dutch Golden Age*. New Haven, CT: Yale University Press, 2007.

Cunningham, Andrew. "The End of the Sacred Ritual of Anatomy." *Canadian Bulletin of Medical History* 18 (2001): 187–204.

Cusatelli, Giorgio, ed. *Viaggi e viaggiatori del Settecento in Emilia e in Romagna*. 2 vols. Bologna: Il Mulino, 1986.

Cushing, Harvey. "Ercole Lelli and His Écorché." *Yale Journal of Biology and Medicine* 9, no. 3 (1937): 199–213.

Dacome, Lucia. "'Un certo e quasi incredibile piacere': Cera e Anatomia nel Settecento." *Intersezioni: Rivista di storia delle idee* 25 (December 2005): 415–36.

———. "Waxworks and the Performance of Anatomy in Mid-18th-century Italy." *Endeavor* 30, no. 1 (March 2006): 29–35.

———. "Women, Wax and Anatomy." *Renaissance Studies* 21, no. 4 (September 2007): 522–50.

Daston, Lorraine, and Peter Galison. *Objectivity*. New York: Zone Books, 2007.

de Carlis, Stefano, and Angelo Turchini. *Giovanni Bianchi: Medico primario di Rimini ed archiatra pontificio*. Villa Verrucchio: Pazzini, 1999.

Didi-Huberman, Georges. "Wax Flesh, Viscous Circles." In Didi-Huberman, von Düring, and Poggesi, *Encyclopaedia Anatomica*, 64–74.

Didi-Huberman, Georges, Monika von Düring, and Marta Poggesi. *Encyclopaedia Anatomica: Museo La Specola Florence*. Cologne: Taschen, 1999.

Fanti, Mario. "Prospero Lambertini arcivescovo di Bologna (1731–1740)." In *Benedetto XIV*, 165–233. Cento: Centro Studi Girolamo Baruffaldi, 1982.

———. "Sulla figura e l'opera di Marcello Oretti." *Il Carrobbio* 8 (1982): 125–43.

Ferrari, Giovanna. "Public Anatomy Lessons and the Carnival: The Anatomy Theatre of Bologna." *Past and Present* 117 (November 1987): 50–106.

Ferretti, Massimo. "Il notomista e il canonico: Significato della polemica sulle cere anatomiche di Ercole Lelli." In *I materiali dell'Istituto delle Scienze*, 100–114. Bologna: CLUEB, 1979.

Findlen, Paula. "Becoming a Scientist: Gender and Knowledge in Eighteenth-Century Italy." In "Scientific Personae," edited by Lorraine Daston and Otto Sebum. Special issue, *Science in Context* 16 (2003): 59–87.

———. "Maria Gaetana Agnesi, Translator's Introduction." In *The Contest for Knowledge*, edited by Rebecca Messbarger and Paula Findlen, 117–27. Chicago: University of Chicago Press, 2005.

———. *Possessing Nature: Museums, Collecting and Scientific Culture in Early Modern Italy*. Berkeley and Los Angeles: University of California Press, 1994.

———. "Science as a Career in Enlightenment Italy: The Strategies of Laura Bassi." *Isis* 84 (1993): 441–69.

———. "Translating the New Science: Women and the Circulation of Knowledge in Enlightenment Italy." *Configurations* 3, no. 2 (1995): 167–206.

———. "Women on the Verge of Science: Aristocratic Women and Knowledge in Early Eighteenth-Century Italy." In *Women, Gender and Enlightenment*, edited by Sarah Knott and Barbara Taylor, 265–87. London: Palgrave Press, 2005.

Findlen, Paula, Wendy Wassing Roworth, and Catherine Sama, eds. *Italy's Eighteenth Century: Gender and Culture in the Age of the Grand Tour*. Stanford, CA: Stanford University Press, 2009.

Focaccia, Miriam, ed. *Anna Morandi Manzolini: Una donna fra arte e scienza; Immagini, documenti, repertorio anatomico*. Florence: Olschki, 2008.

Foucault, Michel. *Discipline and Punish: The Birth of the Prison*. Translated by Alan Sheridan. New York: Vintage Books, 1977.

Fox, Christopher, Roy Porter, and Robert Wokler, eds. *Inventing Human Science: Eighteenth-Century Domains*. Berkeley and Los Angeles: University of California Press, 1995.

Freedberg, David. *The Power of Images: Studies in the History and Theory of Response.* Chicago: University of Chicago Press, 1989.

———. "The Representation of Martyrdoms during the Counter Reformation in Antwerp." *Burlington Magazine* 118 (1976): 128–38.

Garber, Marjorie, and Nancy J. Vickers. *The Medusa Reader.* New York: Routledge, 2003.

Gelbart, Nina Rattner. *The King's Midwife.* Berkeley and Los Angeles: University of California Press, 1998.

Gelmetti, P. "The First Chair in Operative Medicine in Bologna." [In Italian.] *Giornale di batteriologia, virologia ed immunologia ed annali dell'Ospedale Maria Vittoria di Torino* 61, no. 11 (November–December 1968): 546–58.

Giuli, Paola. "Corilla Olimpica Improvvisatrice: Toward a Reappraisal of Her Life." [In English.] In *Corilla Olimpica e la poesia del Settecento: Atti del Convegno Internazionale,* edited by Moreno Fabbri, 155–72. Pistoia: Maschietto e Musolino, 2002.

Gombrich, Ernst H. *Art and Illusion: A Study in the Psychology of Pictorial Representation.* Princeton, NJ: Princeton University Press, 1969.

Griffith, Adrian. "The Foramen of Rivinus—An Artefact." *Journal of Laryngology and Otology* (July 1961): 607–13.

Guerci, Luciano. *La discussione sulla donna nell'Italia del Settecento.* Turin: Tirrenia Stampatori, 1987.

Haynes, Renée. *The Philosopher King: The Humanist Pope Benedict XIV.* London: Weidenfeld and Nicolson, 1970.

Heckscher, William S. *Rembrandt's Anatomy of Dr. Nicolaas Tulp: An Iconographical Study.* New York: New York University Press, 1958.

Hilgartner, Stephen. *Science on Stage: Expert Advice as Public Drama.* Stanford, CA: Stanford University Press, 2000.

Hilloowala, Rumy, Benedetto Lanza, Maria Luisa Azzaroli Puccetti, Marta Poggesi, and Antonio Martelli. *The Anatomical Waxes of the Specola.* Florence: Arnaud, 1995.

Horowitz, Maryanne Cline. "Aristotle and Woman." *Journal of the History of Biology* 9, no. 2 (1976): 183–213.

Hunter, R. H. F. *Physiology of the Graafian Follicle and Ovulation.* Cambridge: Cambridge University Press, 2003.

Jacobs, Fredrika H. *Defining the Renaissance Virtuosa.* Cambridge: Cambridge University Press, 1997.

Jacquart, Danielle, and Claude Thomasset. *Sexuality and Medicine in the Middle Ages.* Princeton, NJ: Princeton University Press, 1985.

Jordanova, Ludmilla. "Sex and Gender." In *Inventing Human Science: Eighteenth-Century Domains,* edited by Christopher Fox, Roy Porter, and Robert Wokler, 152–83. Berkeley and Los Angeles: University of California Press, 1995.

———. *Sexual Visions: Images of Gender in Science and Medicine between the Eighteenth and Twentieth Centuries.* Madison: University of Wisconsin Press, 1989.

Kemp, Martin, and Marina Wallace. *Spectacular Bodies: The Art and Science of Human Bodies from Leonardo to Now.* London: Hayword Gallery Publishing, 2000.

Knoefel, Peter. *Felice Fontana, Life and Works.* Trento: Società di Studi Trentini di Scienze Storiche, 1984.

Lanzarini, Viviana. "Un museo per la didattica e la sanità ostetrica." In *Ars obstetricia bononiensis,* 32–45.

Laqueur, Thomas. *Making Sex: Body and Gender from the Greeks to Freud.* Cambridge, MA: Harvard University Press, 1990.

———. "Sex in the Flesh." *Isis* 94 (2003): 300–306.

Logan, Gabriella Berti. "Women and the Practice and Teaching of Medicine in Bologna in the Eighteenth and Early Nineteenth Centuries." *Bulletin of the History of Medicine* 77, no. 3 (2003): 506–35.

Maerker, Anna Katharina. "The Anatomical Models of the Specola: Production, Uses, and Reception." *Nuncius*: Journal of the History of Science 21, no. 2 (2006): 295–321.

———. "Model Experts: The Production of Anatomical Models at La Specola, Florence and the Josephinum, Vienna, 1775–1814." Ph.D. diss., Cornell University, 2005.

———. "Uses and Publics of the Anatomical Model Collections of La Specola, Florence, and the Josephinum, Vienna, around 1800." In *From Private to Public: Natural Collections and Museums*, ed. Marco Beretta, 81–96. Nantucket, MA: Science History Publications, 2005.

Martinotti, G. *L'Insegnamento dell'anatomia in Bologna Prima del Secolo XIX.* Bologna: Azzoguidi, 1911.

———. *Prospero Lambertini (Benedetto XIV) e lo studio dell'anatomia in Bologna.* Bologna: Tipografica Azzoguidi, 1911.

Masi, Ernesto. "Laura Bassi ed il Voltaire." In *Studi e ritratti*, 157–71. Bologna: Zanichelli, 1881.

Mazzolini, Renato. "Plastic Anatomies and Artificial Dissections," In *Models: The Third Dimension of Science*, edited by Soraya de Chadarevian and Nick Hopwood, 43–70. Stanford, CA: Stanford University Press, 2004.

Mazzolini, Renato, and Giuseppe Ongaro, eds. *Epistolario di Felice Fontana.* Vol. 1, *Carteggio con Leopoldo Marc Antonio Caldani, 1758–1794.* Trento: Società di Studi Trentini di Scienze Storiche, 1980.

Mazzotti, Massimo. "Maria Gaetana Agnesi: Mathematics and the Making of the Catholic Enlightenment." *Isis* 92 (December 2001): 657–83.

———. "Newton for Ladies." Bologna Science Classics Online. http://www.cis.unibo.it/cis13b/bsco3/notbyed2.ASP?id_opera=32&offset=1.

Medici, Michele. *Compendio storico della Scuola Anatomica di Bologna dal renascimento delle scienze e delle lettere a tutto il secolo XVIII con un paragone fra la sua antichità e quella delle scuole di Salerno e di Padova.* Bologna: Tipografia Lelio della Volpe e dei Sassi, 1857.

———. "Elogio d'Ercole Lelli." In *Memorie della Accademia delle Scienze dell'Istituto di Bologna*, 7:158–86. Bologna: San Tommaso d'Aquino, 1856.

———. "Elogio di Giovanni Manzolini e di Anna Morandi, coniugi Manzolini." In *Memorie della Accademia delle Scienze dell'Istituto di Bologna*, 8:3–23. Bologna: San Tommaso d'Aquino, 1857.

Melli, Elio. "Laura Bassi Verati: Ridiscussioni e nuovi spunti." In *Alma mater studiorum*, 74.

Messbarger, Rebecca. "As Who Dare Gaze the Sun: Anna Morandi Manzolini's Wax Anatomies of the Male Reproductive System and Genitalia." In *Italy's Eighteenth Century: Gender and Culture in the Age of the Grand Tour*, edited by Paula Findlen, Wendy Wassing Roworth, and Catherine Sama, 251–71. Stanford, CA: Stanford University Press, 2009.

———. *The Century of Women: Representations of Women in Eighteenth-Century Italian Public Discourse.* Toronto: University of Toronto Press, 2002.

———. "Cognizione corporale: La poetica anatomica di Anna Morandi Manzolini." In

Scienza a due voci, edited by Raffaella Simili, 39–61. Florence: Olschki, 2006.

———. "Re-casting Wax Anatomical Modeler Anna Morandi Manzolini (1714–1774)." In *Dall'origine dei lumi alla rivoluzione: Scritti in onore di Luciano Guerci e Giuseppe Ricuperati*, edited by Donatella Balani, Dino Carpanetto, and Marina Roggero, 353–84. Rome: Edizioni di Storia e Letteratura, 2008.

———. "Re-membering a Body of Work: Anatomist and Anatomical Designer Anna Morandi Manzolini." *Studies in Eighteenth-Century Culture* 32 (2003): 123–54.

———. "Waxing Poetic: Anna Morandi Manzolini's Anatomical Sculptures." *Configurations* 9 (2001): 65–97.

Messbarger, Rebecca, and Paula Findlen. *The Contest for Knowledge: Debates over Women's Learning in Eighteenth-Century Italy*. Chicago: University of Chicago Press, 2005.

Mostra del Settecento Bolognese. Catalogo. Bologna: Mareggiani, 1935.

Muir, Edward. *Civic Ritual in Renaissance Venice*. Princeton, NJ: Princeton University Press, 1981.

Murphy, Caroline P. *Lavinia Fontana: A Painter and Her Patrons in Sixteenth-Century Bologna*. New Haven, CT: Yale University Press, 2003.

Newman, Karen. *Fetal Positions: Individualism, Science, Visuality*. Stanford, CA: Stanford University Press, 1996.

Olmi, Giuseppe, ed. *Rappresentare il corpo: Arte e anatomia da Leonardo all'Illuminismo*. Bologna: Bonobia University Press, 2004.

Ottani, Vittoria, and Gabriella Guiliani Piccari. "L'opera di Anna Morandi Manzolini nella ceroplastica anatomica bolognese." In *Alma mater studiorum*, 81–93.

Pancino, Claudia. "Donne e scienza." In *La memoria di lei*, edited by Gabriella Zarri, 89–101. Turin: Società Editrice Internazionale, 1996.

———. "L'ostetricia del Settecento e la scuola bolognese di Giovanni Antonio Galli." In *Ars obstetricia bononiensis*, 24–31.

Panzanelli, Roberta. *Ephemeral Bodies: Wax Sculpture and the Human Figure*. Los Angeles: Getty Research Center, 2008.

Park, Katharine. "Dissecting the Female Body: From Women's Secrets to the Secrets of Nature." In *Crossing Boundaries: Attending to Early Modern Women*, edited by Jane Donawerth and Adele Seeff, 29–47. Newark: University of Delaware Press, 2000.

———. *Secrets of Women: Gender, Generation, and the Origins of Human Dissection*. New York: Zone Books, 2006.

Park, Katharine, and Robert Nye. "Destiny Is Anatomy." Review of *Making Sex: Body and Gender from the Greeks to Freud*, by Thomas Laqueur. *New Republic*, February 18, 1991, 53–57.

Perini, Giovanna. "La Camera Anatomica dell'Istituto delle Scienze." In *Palazzo Poggi da dimora aristocratica a sede dell'università*, edited by Anna O. Calvina, 176–88. Bologna: Nuova Alfa Editoriale, 1988.

Pinto-Correia, Clara. *The Ovary of Eve: Egg and Sperm and Preformation*. Chicago: University of Chicago Press, 1997.

Plessi, Giuseppe. *Le Insignia degli Anziani del Comune dal 1530–1796 Catalogo-Inventorio*. Rome: Archivio di Stato di Bologna, 1954.

Poggesi, Marta. "The Wax Figure Collection in 'La Specola' in Florence." In Didi-Huberman, von Düring, and Poggesi, *Encyclopaedia Anatomica*, 12–14.

Pomata, Gianna. "Donne e Rivoluzione Scientifica: Verso un nuovo bilancio." In *Corpi e storia: Donne e uomini dal mondo antico all'età contemporanea*, 165–91. Rome: Viella, 2002.

————. "Perché l'uomo è un mammifero: Crisi del paradigma maschile nella medicina di età moderna." In *Genere e mascolinità: Uno sguardo storico,* 133–52. Rome: Bulzoni, 2000.

Pomian, Krzysztof. "Vision and Cognition." In *Picturing Science Producing Art*, edited by Caroline A. Jones and Peter Galison, 211–32. New York: Routledge, 1998.

Porter, Roy. "Medical Science and Human Science in the Enlightenment." In Fox, Porter, and Wokler, *Inventing Human Science*, 53–87.

Praz, Mario. "Le figure di cera in letteratura." In *La ceroplastica nella scienza e nell'arte: Atti del I Congresso internazionale*, June 3–7, 1:549–68. Florence: Olschki, 1975.

Prosperi, Adriano. *Dare l'anima: Storia di un infanticidio*. Torino: Einaudi, 2005.

————. "Esecuzioni captiali e controllo sociale nella prima età moderna." In *La pena di morte nel mondo: Convegno Internazionale*. Bologna, Casale Monferrato, Italy: Marietti, 1983.

————. "Il sangue e l'anima: Ricerche sulle Compagnie di Giustizia in Italia." *Quaderni Storici* 51, no. 17 (1982): 959–99.

Puppi, Lionello. *Torment in Art: Pain, Violence and Martyrdom*. New York: Rizzoli, 1991.

Radcliffe, Walter. *"Milestones in Midwifery" and "The Secret Instrument (The Birth of the Midwifery Forceps)"*. San Francisco: Norman Publishing, 1989.

Ragg, Laura. *The Women Artists of Bologna*. London: Methuen and Co., 1907.

Riccomini, Eugenio. *Mostra della scultura bolognese del Settecento*. Bologna: Tamari, 1966.

Roe, Shirley. *Matter, Life and Generation: 18th-Century Embryology and the Haller-Wolff Debate*. Cambridge: Cambridge University Press, 1981.

Roger, Jacques. *The Life Sciences in Eighteenth Century French Thought*. Edited by Keith R. Benson and translated by Robert Ellrich. Stanford, CA: Stanford University Press, 1997.

Ruestow, E. G. "Images and Ideas: Leeuwenhoek's Perception of the Spermatozoa." *Journal of the History of Biology* 16 (1983): 185–224.

Ruggeri, Alessandro, and Anna Maria Bertoli Barsotti. "The Birth of Waxwork Modelling in Bologna." *Italian Journal of Anatomy and Embryology* 102, no. 2 (April–June 1997): 99–107.

Ruggeri, Franco. "Il Museo dell'Istituto di Anatomia Umana Normale." In *I luoghi del conoscere: I laboratori storici i musei dell'Università di Bologna*, edited by Amilcare Pizzi, 98–105. Bologna: Banca del Monte di Bologna e Ravenna, 1988.

Salvi, Paola. "Da Leonardo alle Accademie: Procedimenti e metodi anatomici degli artisti." In Olmi, *Rappresentare il corpo*, 51–73.

Sappol, Michael. *A Traffic of Dead Bodies*. Princeton, NJ: Princeton University Press, 2002.

Satriani, L. M. Lombardi. "Ex-Voto di Cera in Calabria." In *La ceroplastica nella scienza e nell'arte: Atti del I Congresso internazionale*, 1:533–46. Florence: Olschki, 1975.

Sawday, Jonathon. *The Body Emblazoned: Dissection and the Human Body in Renaissance Culture*. New York: Routledge, 1995.

Schiebinger, Londa. *The Mind Has No Sex? Women in the Origins of Modern Science*. Cambridge, MA: Harvard University Press, 1989.

————. "Skeletons in the Closet: The First Illustrations of the Female Skeleton in Eighteenth-Century Anatomy." In *The Making of the Modern Body*, edited by Catherine Gallagher and Thomas Laqueur, 42–82. Berkeley and Los Angeles: University of California Press, 1987.

————. " Skelettestreit." *Isis* 94 (2003): 307–13.

Schlosser, Julius Ritter von. *Geschichte der Porträtbildnerei in Wachs*. Leipzig: Wein, 1911.

Schlosser, Julius Ritter von, Thomas Medicus, Edouard Pommier, and Gotthold Ephraim Lessing. *Histoire du portrait en cire*. Paris: Macula, 1997.

Schober, Richard. "Gli effetti delle riforme di Maria Teresa sulla Lombardia." In *Economia, istituzioni e cultra in Lombardia nell'età di Maria Teresa*, edited by Aldo de Maddalena, Ettore Rotelli, and Germano Barbarisi, 3:201–14. Milan: Il Mulino, 1982.

Schulz, William F. ed. *The Phenomenon of Torture: Readings and Commentary*. Philadelphia: University of Pennsylvania Press, 2007.

Sheriff, Mary. *The Exceptional Woman: Elisabeth Vigée-Lebrun and the Cultural Politics of Art*. Chicago: University of Chicago Press, 1996.

Sherman, Claire Richter. *Writing on Hands: Memory and Knowledge in Early Modern Europe*. Carlisle, PA: The Trout Gallery, Dickinson College, 2001.

Showalter, Elaine. "The Woman's Case." In *Sexual Anarchy: Gender and Culture at the fin de Siecle*, 127–43. New York: Penguin Books, 1990.

Siraisi, Nancy. *Medicine and the Italian Universities 1250–1600*. Leiden: Brill, 2001.

Smith, Pamela H. *The Body of the Artisan*. Chicago: University of Chicago Press, 2004.

Soppelsa, Maria Laura, and Eva Viani. "Dal newtonianismo per le dame al newtonianismo delle dame: Cristina Roccati una 'savante' del Settecento Veneto." In *Donne, filosofia e cultura nel Seicento*, edited by Pina Totaro, 211–40. Rome: Consiglio Nazionale delle Ricerche, 1999.

Spierenburg, Pieter. *The Spectacle of Suffering: Executions and the Evolution of Repression from a Preindustrial Metropolis to the European Experience*. Cambridge: Cambridge University Press, 1984.

Stafford, Barbara Maria. *Body Criticism: Imaging the Unseen in Enlightenment Art and Medicine*. Cambridge, MA: MIT Press, 1991.

Stolberg, Michael. "A Woman Down to Her Bones: The Anatomy of Sexual Difference in the Sixteenth and Early Seventeenth Centuries." *Isis* 94 (2003): 274–99.

Stoye, John. *Marsigli's Europe, 1680–1730: The Life and Times of Luigi Ferdinando Marsigli, Soldier and Virtuoso*. New Haven, CT: Yale University Press, 1994.

Tega, Walter. "Mens agitat molem: L'Accademia delle Scienze di Bologna (1711–1804)." In *Scienza e letteratura nella cultura italiana del Settecento*, ed. Renzo Cremente and Walter Tega, 65–108. Bologna: Il Mulino, 1984.

————. "Papa Lambertini: Una lucida visione dei rapporti fede-scienza," *Secularia Nona* 13 (1996–97): 92–98.

————, ed. *I commentari dell'Accademia delle Scienze di Bologna*. Vol. 1 of *Anatomie accademiche*. Bologna: Il Mulino, 1986.

Terlinden, Vicomte Ch. "Journal de voyage d'un médicin bruxellois de Munich à Rome en 1755." *Bulletin de l'Institut historique belge de Rome* 23 (1944–46): 123–59.

Terpstra, Nicholas. *Abandoned Children of the Italian Renaissance: Orphan Care in Florence and Bologna*. Baltimore: Johns Hopkins University Press, 2005.

Thompson, Lana. *The Wandering Womb: A Cultural History of Outrageous Beliefs about Women*. Amherst, MA: Prometheus Books, 1999.

Umangulov, Villi. "Sculpture of Anna Manzolini Work by Nollekens in Petergof." *Study Group on Eighteenth-Century Russia, Newsletter* 29 (September 2001): 55–67.

Vassena, Elio. "La fortuna dei ceroplasti bolognesi del Settecento." Ph.D. diss., Università degli Studi di Bologna, 1997.

Weber, John C. *Shearer's Manual of Human Dissection.* 8th ed. New York: McGraw-Hill, 1990.

Weil, Phoebe Dent. "Bozzetti Problems: Philological, Functional Technical." Ph.D. diss., New York University, 1966.

Wilson, Adrian. *The Making of Man Midwifery: Childbirth in England 1660–1770.* Cambridge, MA: Harvard University Press, 1995.

Wilson, Catherine. *The Invisible World: Early Modern Philosophy and the Invention of the Microscope.* Princeton, NJ: Princeton University Press, 1995.

Index